AMINTAPHIL: The Philosophical Foundations of Law and Justice

Volume 10

The *AMINTAPHIL book series* considers the philosophical foundations of law and justice from the perspectives of academic philosophy, practical political science and applied legal studies. The American Section of the International Association for the Philosophy of Law and Social Philosophy ("AMINTAPHIL") supervises the series, which dedicates each volume to the most pressing contemporary problems in legal theory and social justice. AMINTAPHIL holds biennial meetings of leading scholars in philosophy, law, and politics to discuss the philosophical basis of vital questions. The AMINTAPHIL volumes present the ultimate results of these discussions.

More information about this series at http://www.springer.com/series/7372

Joshua J. Kassner

Climate Change and Sovereignty

An Essay on the Moral Nature and Limits of State Sovereignty

Springer

Joshua J. Kassner
Division of Legal, Ethical and Historical Studies
University of Baltimore
Baltimore, MD, USA

ISSN 1873-877X ISSN 2351-9851 (electronic)
AMINTAPHIL: The Philosophical Foundations of Law and Justice
ISBN 978-3-030-73577-7 ISBN 978-3-030-73578-4 (eBook)
https://doi.org/10.1007/978-3-030-73578-4

This Springer imprint is published by the registered company Springer Nature Switzerland AG.
The registered company address is: Gewerbestrasse 11, 6330 Cham, Switzerland

Thank you, Mom, Dad, Jessica, Rebecca, Aaron, and Rachael

Acknowledgements

I have been working on this project for the better part of the past decade. As such, those who have contributed to the development of this book go far beyond those I am able to acknowledge here. With that said, there are several individuals and groups that warrant some special recognition.

To begin, I have bent the ear of Steve Scalet and David Lefkowitz more times than I would like to admit, often subjecting them to the chaos of the debates going on inside my mind. Their willingness to listen and engage with me on any number of ideas and puzzles that I have struggled with has been invaluable, as has their friendships over the past decade.

During the early phases of this project, I was granted a sabbatical year that I spent as an Associate Scholar at the University of East Anglia. While there I was able to present many of the ideas that would eventually become the core philosophical arguments of the book in various fora. In addition, I cannot thank Adriana Sinclair and Alex Brown enough for their thoughtfulness and collegiality while I was there and for their feedback on an early version of this manuscript. This would have been much weaker but for their comments and criticisms.

More recently, I was the Freeman Resident Ethics Fellow at the Vice Admiral James B. Stockdale Center for Ethical Leadership. I want to thank Ed Barrett, David Luban, and Mitt Regan for inviting me to join them for the 2018–2019 academic year. I also want to thank Alycia Laguardia and Brent Kyle, the other fellows for the lively discussions that helped me to formulate many of the ideas in this book.

While a fellow, I was invited by the Department of Philosophy at Utrecht University to present a version of the core arguments of the book at their Practical Philosophy Colloquium. I was met with an audience that was more than willing to offer rigorous criticisms and engage in vigorous philosophical debate. In the end, this book has only benefited from that experience.

I would like to thank my colleagues at the University of Baltimore who have helped to foster my academic ambitions for the past 15 years. I would also like to acknowledge the patience and assistance that has been provided by numerous individuals at Springer. And, I want to thank the anonymous reviewer whose comments helped me put the finishing touches on this project.

Lastly, on a personal note, I want to thank Stephanie, Mackenzie, and Abigail. My wife and daughters are an unending font of love, joy, and support. An ever-present reminder of why this kind of work is valuable.

Contents

Chapter 1
Introduction and Overview

> Westphalia, for better or worse, marks the end of an epoch and the opening of another. It represents the majestic portal which leads the old into the new.[1]

In 1648, after decades of internecine war, through their execution of the Treaties of Osnabruck and Munster,[2] the powers of Europe sought to bring an end to the violence that had ravaged their kingdoms. This moment in the history of international relations has come to be known as the Peace of Westphalia, and it is here that we find the intellectual foundations for the modern conception of sovereignty. To be specific, the Pea8e of Westphalia introduced territorially defined sovereign jurisdiction as a means of promoting the peace between and security of independent political communities.[3]

Much has changed, however, since Westphalia with some of the most dramatic changes occuring during the latter half of the twentieth and opening decades of the twenty-first centuries. Economic globalization, the industrialization of the developing world, the communications revolution, the growing importance of human rights in international and domestic affairs, the rise of non-state actors, and anthropogenic climate change are just a few of the most impactful.

These changes have brought about a world in which the significance of geographic distance and political boundaries to the ways in which individuals and groups relate to one another has been greatly reduced. Individuals, peoples, corporations and states are becoming ever more economically, environmentally, culturally, politically and epistemically interdependent. New moral and practical challenges faced by the international community (e.g. climate change, global terrorism, immigration, etc.) have emerged as a result of this growing interdependence.

Westphalian sovereignty, with its focus on international peace and security, seems ill-suited for such a pervasively interdependent environment. In fact, when

[1] Gross (1948), pp. 20, 28.

[2] Treaties of Osnabruck and Munster (Jan 30, 1648).

[3] *Ibid.*

J. J. Kassner, *Climate Change and Sovereignty*, AMINTAPHIL: The Philosophical Foundations of Law and Justice 10, https://doi.org/10.1007/978-3-030-73578-4_1

invoked, sovereignty often stands as an obstacle to the international community's ability to respond to the new and emerging challenges brought about by such interdependence. This apparent conflict between respect for state sovereignty and the moral and practical demand that these new challenges be addressed has caused many to question whether state sovereignty remains relevant to the normative framework of international relations, and whether the role it plays in that normative framework can be justified.

Despite this, and somewhat surprising given the significance of sovereignty to contemporary international law and relations, there has been scant rigorous philosophical inquiry focused on state sovereignty itself.[4] This book, is in part, an effort to fill this gap. I begin by showing that the accepted understanding of the moral nature and limits of state sovereignty is wanting. I then develop an understanding of the moral nature and limits of state sovereignty based on an exploration of the implicit limitations on, and underlying moral bases for, the Westphalian account of sovereignty.

Finally, to connect theory to practice, I will assess the implications the account of the moral nature and limits of state sovereignty developed here has for the challenge of climate change, assessing what role sovereignty ought to play in the practical deliberations of states as they decide whether and how to respond to the challenge of climate change, including some suggestions for institutional design and reform.

This introductory chapter serves several valuable purposes. It is a brief statement on the conception of sovereignty at issue. It provides an explanation of the impetus behind this project and its theoretical and practical importance. By addressing various preliminary matters relevant to the broader inquiry, it serves an important methodological purpose by setting the stage for the inquiry that follows. Then, finally, it provides an overview of the arguments and analyses that make up this book.

1.1 Brief Statement on Sovereignty

Whether one is discussing the permissibility of humanitarian intervention, the right of a group to secede, or the responsibility of the international community to respond to issues like global climate change or human trafficking, determining what ought to be done is dependent on understanding the role state sovereignty plays, or ought to play, in the practical deliberations of states and the normative framework of international relations. As such, and though a more detailed account follows in Chap. 2, it is important to have a working understanding of the notion of sovereignty at issue.

Different conceptions of sovereignty are employed for different purposes in a variety of distinct but related scholarly and institutional contexts.[5] Political and

---

[4]Bartlelson (1995), p. 5.

[5]Karp (2008), pp. 313–335.

moral philosophers are concerned with questions of justification and legitimacy.[6] Legal scholars seek to understand sovereignty as a juridical norm.[7] Political theorists pursue an understanding of the relevance (or irrelevance) of sovereignty in the strategic interactions between states.[8]

Some challenging the normative understanding of the principle itself advocate for an understanding based in realism or rational choice theory and argue that appeals to sovereignty are merely pretext or the epiphenomena of states pursuing their national interests.[9] In light of the practical importance of sovereignty and the potential for misunderstandings that could arise from the conceptual ambiguity surrounding the concept itself, it is essential that one understand the particular notion at issue.

The understanding of sovereignty with which this project is concerned, is sovereignty as a norm grounded in our shared political history and governing the interactions of states in the international arena. As such, it makes sense to start with a brief discussion of its origins. In doing so, it may make sense to start with a cursory understanding of the European political world before the Peace of Westphalia. Pre-Westphalian political relationships were built upon patronage, personal allegiance, and religion.[10] The political structure this created was a significant cause of the wars that devastated Europe prior to Westphalia. Thus, the motivation underlying the Peace of Westphalia is not difficult to understand. As Mark W. Janis has noted, "Europeans sought a simpler, and it was hoped, safer set of loyalties."[11]

After Westphalia, a new political system emerged, one that demanded respect for the independence of territorially defined political communities. Initially, with the goal of promoting the peace between and security of those political communities, this meant that political communities respect one another's right to make choices regarding religion and other domestic affairs.[12] Over time the chaotic political relations of the pre-Westphalian world were transformed into what Michael Smith has called "international order".[13] This commitment to protect and promote international peace and security lies at the heart of our contemporary understanding of the value of sovereignty, is a cornerstone of the modern state system and is stated explicitly in the Charter of the United Nations.[14]

In addition to international peace and security, Hans Morgenthau has argued that the French Revolution marked the emergence of another important purpose served by sovereignty. Namely, Morgenthau notes that the leaders of the French Revolution

[6]See Luban (1980), pp. 392–397. See also Shue 1996 And see Walzer (1985), pp. 217–237.

[7]See Janis (1999). See also Byers (1999). And see Cohen (2012).

[8]See Krasner (1999). See also Karp (2008). And see Sinclair (2010).

[9]See Goldsmith and Posner (2005); and see Sinclair (2010).

[10]Morris (1998).

[11]Janis (1999), p. 159.

[12]Janis (1999), pp. 159–160.

[13]Smith (1999), p. 274.

[14]United Nations Charter, Ch. 1 Art. 1 Section 2.

claimed that foreign powers were required to stay out of the internal political affairs of France, not as a prophylactic against disorder, but as a recognition of France's right to political self-determination.[15] Thus, sovereignty was also argued to protect each political community's right to self-determination.[16]

During the twentieth century, this more inwardly focused understanding of the value of state sovereignty was (and continues to be) embraced by the international community. It is held in particularly high regard by states emerging from the shadow of occupation and colonialism. Consequently, in addition to its role in the promotion of international order, sovereignty's value in relation to political self-determination continues to be an essential commitment of the normative framework of international relations.[17]

It is important, however, to recognize that sovereignty has not remained static. For example, the Universal Declaration of Human Rights (UDHR) was constructed as a response to some of the worst atrocities of the Second World War. Though aspirational in nature, the substantive commitments of the UDHR were made manifest through a series of conventions aiming to give legal effect to the hopes of the UDHR.[18] More recently, in 2005, the United Nations General Assembly endorsed the International Commission on Intervention and State Sovereignty's (ICISS) effort to reconcile Westphalian sovereignty with the international community's commitment to human rights through the Responsibility to Protect (RtoP) doctrine.[19] Under RtoP, legitimate claims to sovereignty depend on the willingness and ability of a state to protect the human rights of its inhabitants.

Nonetheless, the evolution of the principle has not significantly altered the underlying objectives and values sovereignty is taken to serve.[20] Even RtoP is based on the proposition that international peace and security is served by protecting human rights.[21] As such, international peace and security and political self-determination remain the primary bases for understanding the value of sovereignty within the normative framework of international relations. Respect for sovereignty is

[15] *Ibid.*

[16] Morgenthau (1967), pp. 425–436.

[17] United Nations Charter, Ch. 1 Art. 1 Section 2.

[18] International Convention on the Elimination of All forms of Racial Discrimination (Dec 21, 1965); International Covenant on Civil and Political Rights (Dec 16, 1966); International Covenant on Economic, Social, and Cultural Rights (Dec 16, 1966); Convention on the Elimination of All forms of Discrimination Against Women (Dec 18, 1979); Convention against Torture and Other Cruel, Inhuman or Degrading Treatment or Punishment (Dec 10, 1984); Convention on the Rights of the Child (Nov 20, 1989); International Convention on the Protection of the Rights of All Migrant Workers and Members of Their Families (Dec 18, 1990); and, International Convention for the Protection of All Persons from Enforced Disappearance (Dec 20, 2006).

[19] United Nations World Summit Outcome Document (2005).

[20] Jackson (2007), pp. 2, 23.

[21] International Commission on Intervention and State Sovereignty, Summary Report (2001).

taken to be an essential ***means*** to achieving and maintaining international order[22] and to ensuring that political communities can exercise political self-determination.

From this, two aspects essential to understanding sovereignty can be discerned. First, sovereign states are rights-holders and duty-bearers within a community of rights-holders and duty-bearers. Specifically, sovereign states hold rights against other states and the international community, and bear the duties correlative to the same rights held by other sovereign states and potentially owed to others with moral standing within the international moral community. It is this understanding of sovereignty that enables it to serve the international community's pursuit of international order and protection of political self-determination. Second, the valuable purposes state sovereignty has historically served inform our understanding of the content of those rights and duties and the nature of the underlying moral justification for sovereignty so understood.

An additional aspect of sovereignty not captured by the discussion thus far is the role sovereignty plays in the normative framework of international relations. Understanding that role depends on the role sovereignty plays in the practical deliberations of states and the international community. This, in turn, requires that one understand the nature of the practical reasons to which sovereignty gives rise. A sovereign state is a right-holder; and, provided the matter under consideration falls within the extension of a sovereign state's rights, the state has discretion over whether and/or how to respond to the matter at hand. This discretion is largely content-neutral, meaning that the right-holding state is permitted to engage in many actions that might normally be considered morally objectionable if considered on their own merits.

It is this conception of sovereignty as the moral standing of states in the international arena with which this inquiry is concerned. From this point forward, unless otherwise indicated "sovereignty" refers to this notion. This brief discussion of state sovereignty is no doubt incomplete, but as noted above, Chap. 2 offers a more detailed explanation of the particular notion at issue.

1.2 Impetus and Importance

For many, the necessity of this inquiry may be clear. This is a sentiment, however, that may not be universal. As such, it may be helpful to provide a brief comment on the impetus and importance of this project. This can be understood both as a matter of scholarly importance in relation to international normative political theory and as a matter of the practical policy choices faced by the international community.

---

[22]Janis (1999), p. 159.

1.2.1 Impetus

The underlying motivation for this project began with a recognition of the historical fact, previously discussed, that the nascent idea of what is now the modern mainstream notion of sovereignty was introduced as a reaction to particular problems faced by Europe in the seventeenth century. The changes that have occurred since Westphalia, however, have fundamentally transformed the way in which individuals, peoples, corporations and states relate to one another.

Because of these changes our world is evermore interconnected and interdependent in a variety of practically and morally significant ways (e.g., economically, environmentally, culturally, epistemically, and politically). From the protection of human rights to the moral and practical demand to address climate change, these changes and this greater interconnectedness have given rise to a wide range of new and emerging moral and practical challenges facing the international community.

Despite the transformational nature of these changes and the new challenges they have wrought, sovereignty remains the core commitment of the normative framework of international relations. And, whether one is concerned with growing global inequality, the protection of the human rights of LGBT individuals around the world, the regulation of greenhouse gases in the United States and China, or responding to various security threats arising from non-state actors, sovereignty often stands as an obstacle to efforts to address these challenges. This apparent tension between sovereignty and the moral and practical imperative that these new and emerging challenges be addressed has caused many to question the continued relevance of and justification for sovereignty within the normative framework of international relations.[23]

For some, we are as a matter of empirical fact,[24] or ought as a matter of moral or practical prescription to be, moving into a post-sovereign world. Others contend that the best way to respond to these new challenges is to construct political and legal institutions that strengthen sovereignty.[25] Still others argue that sovereignty remains and ought to remain the foundation of the normative framework of international relations, but that states should voluntarily cede portions of their sovereignty to transnational and global institutions to address these new challenges.[26] Finally, there are those who offer reconstructed understandings of sovereignty, seeking to reconcile the Westphalian notion with the moral demand that these new and emerging challenges be addressed.[27]

This inquiry was motivated, at least in part, by an interest in this debate over the relevance of and justification for sovereignty. The discussion that follows is not,

[23]Steger (2003), p. 61.

[24]MacCormick (1996), pp. 553–567.

[25]Chertoff (2009). And see Rabkin (1998).

[26]Singer (2004).

[27]ICISS (2001).

however, intended to stake out a position in ***this*** debate. Rather, one of its primary objectives, and one of the main reasons for pursuing this project, is based on the proposition that this debate is based on a mistaken assumption about the moral nature and limits of state sovereignty.

To clarify, though differing in their conclusions regarding the relevance and justifiability of sovereignty, these competing accounts all begin by recognizing that there is a moral conflict between respecting sovereignty and satisfying the moral and practical demand that the new and emerging challenges facing the international community be addressed. The debate is, in large part, about which side of this conflict we should assign moral priority—respect for state sovereignty ***or*** meeting the moral and practical demands generated by these challenges.

The conflict only arises, however, if we assume that the new and emerging challenges facing the international community fall within the legitimate zone of moral discretion held by sovereign states. More to the point, assuming that the matter under consideration falls within a state's legitimate zone of sovereign discretion implies that, as an implication of sovereignty, a sovereign state is to be free from outside interference with its choices and actions even if those actions are morally objectionable.[28]

An example may help to explain the nature and significance of this shared assumption. Taking climate change as an example, if climate change falls within a state's sovereign discretion, then ***whether*** and ***how*** each state responds to the challenge of climate change is a matter for the state—and the state alone—to decide. The state may decide that climate change isn't anthropogenic and do nothing. Or, the state may decide that, even though climate change is anthropogenic, that it will still do nothing. Or, the state may simply ignore the problem altogether, and go about its business as usual.

Assuming, for the sake of argument, that climate change falls within the moral discretion of a sovereign state, no matter what the state chooses to do, the state's permission is not dependent on its choice being correct or reasonable. In fact, we can assume that the state's choice is demonstrably wrong—for example, the state may choose to do nothing because the state has concluded that climate change isn't anthropogenic. However, like an individual's right to dispose of their property as they see fit, sovereign discretion permits the right-holder to engage in morally objectionable actions free from interference by others. Thus, much depends on what constitutes the moral nature and limits of state sovereignty.

In the discussion that follows, I argue that this shared assumption about sovereign discretion is mistaken. Consequently, the current debate about the relevance and justifiability of state sovereignty is misguided due to the fact that it is founded upon an illicit assumption. Recognizing this fact opens the conceptual space for an independent investigation into the moral nature and limits of state sovereignty. Which is, in large part, a matter of determining the proper normative breadth of the legitimate zone of moral discretion held by sovereign states.

---

[28]See Shue (2004). And see Kratochwil (1989), pp. 26, 33.

1.2.2 Importance

As to the importance of this book, it can be understood from two distinct perspectives:

1. As a work of political and moral philosophy, it is a contribution to the philosophical understanding of the norms and principles governing the normative framework of international relations; and
2. As a practical matter, it offers an understanding of and justification for the role sovereignty ought to play in the normative framework of international relations.

As to (1.), the fact that there has been little rigorous philosophical analysis of the moral nature and limits of state sovereignty leaves a significant gap in our understanding of one of the most important normative principles governing international relations. The inquiry pursued in this book should help to fill this gap and shape the moral conversation about sovereignty going forward.

It is worth noting that this is not to say that there hasn't been anything written on sovereignty. There have been other scholarly works in which discussion of sovereignty as a moral norm was relevant. To be specific, the projects from which this one is to be distinguished fall into two separate but often overlapping categories. First, are those inquiries focused on specific issues of global moral concern, e.g., humanitarian intervention, human rights, and political self-determination. The aim of such inquiries is a critical moral appraisal of the specific issue under consideration. There, sovereignty is discussed to the extent that it plays a role in the effort to understand the moral issue in question.

There are at least two reasons for distinguishing this project from those that fall into this first category. For one, such inquiries are concerned with the moral analysis of individual substantive issues. Consequently, due to the differences to be found in the relevant moral and circumstantial properties of each issue, the analyses and arguments relevant to our understanding of each issue will be different. Additionally, this project is focused on assessing the moral nature and limits of sovereignty as a normative principle; in other words, this book is ***about sovereignty*** rather than about some substantive moral issue in which sovereignty plays some role.

As to the second category, it is comprised of those projects focused on more general normative issues, e.g., questions of legitimacy, authority, and justice. The treatment of sovereignty in such projects is largely confined to assessing the degree to which the conventional understanding of sovereignty measures up to the implications of some free-standing account of justice or legitimacy. For example, one might defend a cosmopolitan account of global justice and then assess possible justifications for the conventional understanding of state sovereignty based on that account of justice.

The difference between this project and those can be understood by invoking a distinction highlighted by Amartya Sen.[29] There are two approaches to the moral

[29]Sen (2009).

analysis of a norm.[30] One abstracts away from the current practices and institutions in which the norm plays a role, seeking to identify an ideal. The other is grounded in such practices and institutions, and is, what Sen calls, "realization focused."[31] As Sen notes, "[i]mportance must be attached to the starting point, in particular the selection of some questions to be answered (for example, 'how would justice be advanced?'), rather than others (for example, 'what would be perfectly just institutions?')."[32] The work of Gillian Brock and others fall within the latter category and this project in the former.[33]

To help clarify the significance of this distinction, as a theoretical matter, those projects seeking to transcend existing practices and institutions do not treat sovereignty as a matter of direct conceptual concern. Rather, sovereignty becomes the object of critique after a free-standing account of justice or legitimacy has been developed. Here, however, sovereignty is the primary object of conceptual analysis. The basis for critique is not a free-standing (or idealized) account of justice, but the best moral understanding of, or justification for, state sovereignty as it operates within the normative framework of international relations.

Since both projects are focused on the critical analysis of state sovereignty, one might argue that the difference between these inquiries is less significant than I am suggesting. Admittedly, both are concerned with how the world ***ought*** to operate, and either approach can generate prescriptions. The significance of the difference is apparent, however, when one considers the nature of the prescriptions generated by either approach. Prescriptions drawn from free-standing accounts of justice are likely to be more utopian, revolutions not reformation. On the other hand, prescriptions drawn from accounts focused on the moral bases for actual practices and institutions are likely to be oriented towards reform.

As to the practical implications of this project, one should not assume that by adopting the more reform-oriented approach the arguments and conclusions defended are conservative of the *status quo*. In fact, the conclusions reached, and prescriptions defended are radical and far-reaching. For example, I argue that sovereignty does not include moral discretion to decide whether and how to respond to climate change, and that agreements like the Paris Climate Agreement ought to be considered illegitimate. The fact that such dramatic conclusions can be drawn from such weak premises speaks to the strength of the account developed here.

Also speaking to the importance of this project, sovereignty continues to serve as the foundation for the normative framework of international relations. The discussion that follows will help to frame our thinking about the legitimacy of institutions intended to address new and emerging challenges facing the international community. In other words, there are significant implications for how the institutions and

[30]*Ibid*, p. 7.

[31]*Ibid*.

[32]*Ibid*, p. 9.

[33]Brock (2009).

practices that make up the normative framework of international relations ought to be designed and/or reformed.

Lastly, one of the most valuable functions of political philosophy is laying bare the naked power and interests, often cloaked in the rhetoric of rights and respect, leading a community—in this case an international one—down a dangerous path. With this in mind, perhaps the most impactful practical implications of this book, are the ones it has for the rhetorical use of sovereignty within the international arena. Despite the significance of such appeals, at present we lack the conceptual and moral understanding necessary to have a principled basis for distinguishing between legitimate and illegitimate claims to the protections of sovereignty. With greater clarity and understanding we will be better able to evaluate the rhetorical use of sovereignty and hold our leaders accountable for the choices they make on our behalf.

1.3 Preliminary Matters: Setting the Stage

Before turning to the project overview, there are several preliminary matters that need to be addressed. For one thing, there are skeptical concerns about the project that should be acknowledged and to which some response is owed. Additionally, in order to avoid unnecessary confusion and to aid in one's understanding of the arguments and analyses that follow, it is essential that one have a more thorough understanding of the inquiry being pursued.

1.3.1 Some Skeptical Concerns

The first skeptical concern is grounded in a realist perspective on international relations. In short, some realists contend that, like other moral and legal norms, sovereignty is practically irrelevant to understanding the relations between and actions of states in the international arena.[34] This claim is based on the proposition that states are guided, not by normative principles like sovereignty, but by national interest alone. Any appearance of rule following is little more than epiphenomena, a side-effect of the rational pursuit of national interest.

Though I understand the significance of this challenge, I will not offer a detailed response here for three reasons. First, I have addressed the underlying realist proposition elsewhere, and there I concluded that the premises upon which such skepticism depends are unjustified.[35] Second, the main points of the debate have been exhaustively covered by others, and I have little to add other than what I have

[34] Goldsmith and Posner (2005).

[35] Kassner (2012), pp. 12–18.

already contributed. Third, and perhaps most relevant to the nature of this inquiry, is that sovereignty is invoked by world leaders and policymakers as a distinctly ***moral*** justification for the choices they make and actions they take in the international arena, as well as a norm they rely on to ground the expectations they have for the behavior other states. Evaluating such invocations requires that one begin with a charitable understanding of the intentions of those appealing to sovereignty. In other words, rather than presume appeals to sovereignty are a ruse, we should, at least initially, take such rhetoric seriously.

Relatedly, and somewhat counterintuitively, by providing a standard that can be used to judge the legitimacy of appeals to sovereignty, I may, in the end, provide a theoretical benefit to those who are skeptical of such norms and the rhetoric employing them. If it turns out that the standard developed here, when used to assess such rhetoric, implies that appeals to sovereignty are unjustified, then this is a conceptual victory for the skeptic.

One's skepticism might begin, not with the realists' concerns about the practical relevance of legal and moral norms, but with a concern over the choice to ground this inquiry in a critical perspective that begins with the best moral justification for the extant conventional understanding of state sovereignty. More to the point, one might argue that we ***ought*** to engage in ideal theory, in a direct critique of sovereignty based on a free-standing account of justice or legitimacy. Along these lines, one might argue that if state sovereignty proves incompatible with our best free-standing account of justice or legitimacy, then efforts to understand its nature and limits are moot.

We should be wary of this objection. First because it is based on the proposition that these are mutually exclusive paths of inquiry; that in choosing one we are implicitly rejecting the salience and significance of the other. There seems to be little reason to believe in the truth of this proposition. They are different inquiries, and preference for one over the other will depend on the broader purpose of the inquiry being pursued.

Second, one of the primary objectives of this project is to engage those who continue to invoke and depend on Westphalian sovereignty. Whether appealed to by theorists, diplomats, political leaders, judicial officials or journalists, sovereignty continues to play a significant role in our understanding of how international relations works, framing how we think about the normative relations between states. As one of the main objectives of this project is to engage with those who continue to rely on sovereignty so understood, their commitment ought to be taken seriously. As such, there is little to be gained by disregarding the very understandings upon which so many rely.

Finally, there are several values served by the rigorous conceptual and moral analysis of conventional norms. For one, there is the value of having a clearer and more accurate understanding of the concepts on which our extant practices rely. There is also the contribution such an understanding would make to more practical concerns. For example, such an understanding aids in our ability to identify what role a norm ought to play in the practical deliberations of those agents who claim to be guided by it, allowing us to distinguish between legitimate and illegitimate invocations of the norm. This, though valuable in its own right, also serves other important political values like transparency and accountability.

1.3.2 Understanding the Inquiry

In light of this complex conceptual landscape, avoiding unnecessary misunderstandings also requires that one understand the inquiry being pursued. To begin, it is necessary to distinguish between normative and empirical inquiries.[36] This may seem obvious, but the line dividing normative analyses from empirical is often blurred. A typical empirical project might involve explaining how sovereignty as a conventional norm in fact operates. Such inquiries would, however, also have normative implications for actors within the international arena. For example, in circumstances where the actions of different actors are interdependent, as is the case in the international arena, what one actor ***ought*** to do as a prudential matter is dependent on the actions of others.

There are, however, significant methodological differences between empirical inquiries seeking to understand the implications of a conventional norm and inquiries into the role the norm ought, as a matter of moral justification, play in a conventional normative system. The former begins with an account of how the norm actually works in the conventional system. From this, one can predict the behavior of others within the system and, based on these predictions, offer ***practical*** prescriptions for how an agent ought to act to achieve their goals given the likely actions of others. At no point need one assess the underlying moral justification for the conventional understanding of the norm.

Here, on the other hand, the concern is with identifying the best moral justification for the conventional understanding of sovereignty; and then, based on that justification, assessing what role—as a moral matter—sovereignty ought to play in the normative framework of international relations. The empirical understanding of the extant conventional practice is, ultimately, the object of analysis and critique. Once we have an account of the best moral justification for the conventional understanding of sovereignty, we can then derive specific ***moral*** prescriptions that may (or may not) be supportive of particular aspects of our practice.

Identifying the inquiry as normative and moral rather than empirical and prudential leaves certain questions about the normative nature of the inquiry unaddressed. To clarify, it is normative in three respects. First, it is a conceptual analysis of a norm, wherein, an explication of the underlying moral justification for the conventional understanding of sovereignty is offered. Second, in providing an account of the role sovereignty ought to play in the normative framework of international relations, it is normative in a general prescriptive sense, providing moral guidance to actors within the international arena. Third, and finally, the role sovereignty ought

---

[36] Karp (2008).

to play with regards to particular substantive concerns is also addressed through the assessment of whether the challenge of climate change falls (or ought to fall) within the legitimate zone of sovereign discretion. As such, it is normative in a specific and applied sense as well.

One other way to think about the underlying inquiry that may provide further clarity is succinctly captured by David Jason Karp. As he puts it, "[m]any possible decisions fall inside of the boundaries of 'legitimate state action', while others fall outside of them. It is these *boundaries* that are the subjects of a single normative discussion about sovereignty, which focuses on whether the boundaries are, or are not, appropriately placed."[37]

Determining the proper placement of such boundaries is inexorably linked to our understanding of the moral nature and limits of state sovereignty. First, one must account for the fact that sovereign states are rights-holders and duty-bearers within a community of rights-holders and duty-bearers. Because the rights held by each state imply correlative duties borne by other states, the rights and their correlative duties are limits on the legitimate zone of sovereign discretion and, by implication, mark out a moral limit to state sovereignty. Second, one must identify the underlying justification for the moral discretion afforded sovereign states, as the moral justification for sovereignty will carry with it limitations on the legitimate zone of sovereign discretion.

Together, these two considerations imply a third. Namely, if we begin with the assumption that the challenges facing the international community are different in morally and practically significant ways, then whether any particular challenge falls (or ought to fall) within the legitimate zone of sovereign discretion must be determined on a case-by-case basis. Since understanding the moral limits of state sovereignty is dependent on understanding the limits to the legitimate zone of sovereign discretion, determining the moral limits of state sovereignty requires a piecemeal assessment of whether specific challenges facing the international community fall (or ought to fall) within the legitimate zone of sovereign discretion.

An example may help to clarify this point. Since the UDHR, the provision and protection of human rights has become a central aim of the international community. But does the provision and protection of human rights fall within the legitimate zone of sovereign discretion? This question cannot be answered in the abstract. Rather, it must first be the case that a state's choices regarding whether and how to respond to the demand that human rights be provided and protected do not violate any duties owed to others within the international moral community.

Next, even if there is not an inherent conflict between the state's choice of action and the duties it owes to others, it must also be the case that the underlying moral justification for such discretion can accommodate the provision and protection of human rights within such boundaries. In short, the moral bases justifying the discretion afforded sovereign states must justify discretion over the provision and protection of human rights. Each challenge facing the international community must

[37] Karp (2008).

be similarly assessed to determine whether it falls (or ought to fall) within the legitimate zone of sovereign discretion. Identifying the moral limits to sovereignty and sovereign discretion is an ever-evolving process in which our present understanding is inescapably provisional and subject to revision. Categorical claims about the breadth of such discretion based on an abstract understanding of sovereignty and sovereign discretion cannot be sustained.

1.4 Project Overview

The following is a brief overview of the arguments and analyses that make up this book. As noted, this project has both theoretical (the moral nature and limits of state sovereignty) and practical (the role state sovereignty ought to play in the normative framework of international relations) objectives. As such, the book is divided into two sections. *SECTION I* (Chaps. 2–5) addresses the moral nature and limits of state sovereignty. Then, in an effort to connect theory to practice, the answer developed in *SECTION I* is then applied in *SECTION II* (Chaps. 6–11) to the challenge of climate change, determining whether the challenge of climate change falls (or ought to fall) within the legitimate zone of discretion, culminating in suggestions for institutional design and reform.

Though they will be dealt with separately, the theoretical and practical aspects of this project are related. Assessing the role state sovereignty ought to play in the normative framework of international relations requires that one understand the moral nature and limits of state sovereignty. Yet, since the limits of sovereign discretion can only be ascertained in a piecemeal fashion, a complete understanding of the moral limits of sovereignty depends on a case-by-case assessment of the various new and emerging challenges facing the international community.

1.4.1 Section I: The Moral Nature and Limits of State Sovereignty

In *Section I,* the first task to be accomplished is grounded in the proposition that this discussion is premised on the rejection of a tension thought to exist between respect for state sovereignty and the moral and practical demands generated by the new and emerging challenges facing the international community. This (perceived) tension is based on a specific understanding of the relationship between the following elements:

1. The notion of state sovereignty at issue;
2. The understanding of the legitimate zone of sovereign discretion implicit in that notion;

3. The role sovereignty plays in the practical deliberations of states and the international community; and
4. The dramatic nature of the changes that have given rise to the new challenges facing the global community.

Chapter 2 provides a better understanding of this tension, wherein it is explained how the perceived tension depends on a particular understanding of each of the elements noted and how these understandings lead to the perceived tension. In addition, and crucially, Chap. 2 presents an argument that (2.) is based on an unjustified assumption about the legitimate zone of sovereign discretion. If correct, this opens the conceptual space for a renewed inquiry into the moral nature and limits of state sovereignty.

Next, having opened the conceptual space for an inquiry into the moral nature and limits of state sovereignty, Chaps. 3 and 4 take on that task, following two distinct lines of inquiry. Chapter 3 does not engage in an analysis of the moral nature of state sovereignty *per se,* rather it assesses what is implied when one understands sovereignty as being about the standing of states as rights-holders and duty-bearers within a moral community of rights-holders and duty-bearers. Each state's rights are limited by the rights held by, and correlative duties owed to, other rights-holders within the international moral community.

Then, Chap. 4 critically assesses alternative moral justifications for sovereignty, beginning with the identification of the conditions any account seeking to justify state sovereignty must satisfy. With those conditions in hand, the argument in Chap. 4 demonstrates that the alternatives either fail to satisfy those conditions or, if they can satisfy the conditions identified, lead to deeply counterintuitive implications. Of particular importance is the fact that sovereignty is presumed to act as a second-order exclusionary reason,[38] a notion discussed in more detail in Chap. 2. In the end, the best available moral justification for sovereignty is one grounded in the instrumental moral value of sovereignty. Chapter 4 ends with an assessment of the implications this understanding of the moral nature of sovereignty and its justification has for the legitimate zone of sovereign discretion.

Though it will be discussed in greater detail in Chap. 4, the defense of the instrumental justification for state sovereignty begins with an argument tied to the fact that sovereignty is a political artifact created for particular purposes. This is not a philosophical defense of the instrumental account, but it provides good reason for placing the burden of proof on those who would claim that the value of sovereignty is in some way non-instrumental. After making this methodological point, it is necessary to explain what it means for sovereignty to be instrumentally valuable, demonstrate how it satisfies the necessary conditions, and explain the advantages such an account has for how we think about sovereignty in an increasingly interconnected and interdependent world.

---

[38]Raz (1999), pp. 35–42.

The connection between the underlying moral justification and the limits this implies for the legitimate zone of sovereign discretion may not, at first, be apparent; however, what falls (or ought to fall) within the legitimate zone of sovereign discretion depends on what the underlying justification can accommodate. If the best justification for sovereignty is one that relies on its inherent value, then the calculus we would use to determine whether a challenge is within the legitimate zone of sovereign discretion would be grounded in respecting the inherent value of sovereignty. On the other hand, if sovereignty is justified by its instrumental value, then whether we ought to include the challenge at hand within the legitimate zone of sovereign discretion will depend on whether such inclusion is instrumentally valuable to our efforts to respond to the challenge under consideration.

Thus, there are two bases for our understanding of the limits to sovereign discretion. One grounded in those limits inherent to the notion of sovereignty itself and the other tied to the instrumental justification for sovereignty. One might wonder why both tacks need to be pursued. There are at least two reasons for this approach. First, there is value in being accurate about our concepts. One does not have a complete understanding of the moral nature and limits of state sovereignty if either is ignored.

The other reason is more practical in nature and is tied to the inquiry pursued in *Section II*. In some cases, applying either of the two tacks to determine whether a challenge falls (or ought to fall) within the legitimate zone of sovereign discretion will provide similar answers. For others, however, if one were to focus on the first line of argument alone this might result in the conclusion that the matter under consideration is within a state's legitimate zone of sovereign discretion. Such conclusions, however, would be based on only a partial understanding of the moral limits of state sovereignty and could lead to unjustified conclusions. As such, the second tack—one based in an understanding of the underlying moral justification for sovereignty—is essential if we are to have a thorough understanding of what role state sovereignty ***ought*** to play in the practical deliberations of states.

Having defended two distinct limitations on the legitimate zone of sovereign discretion, Chap. 5 explains what implications these limitations have for the practical deliberations of states. It will be important to identify how the limitations implied by the account defended differ from the current understanding. This will be accomplished in the final step of *Section I*, not by challenging the exclusionary effect state sovereignty is presumed to have on the practical deliberations of states, but by demonstrating how the limitations defended have a significant impact on the extension of the exclusionary effect. Finally, Chap. 5 culminates with the development of a framework for assessing whether a challenge falls (or ought to fall) within the legitimate zone of sovereign discretion.

1.4.2 Section II: The Transition from Theory to Practice

Section II makes the transition from theory to practice. It includes the development and application of the evaluative framework provided at the end of *Section I*. However, before outlining the argumentative approach in *Section II*, it is important to explain the purposes *Section II* serves. First, in explicating the framework, the reader should gain insight into how we ought to think about challenges facing the international community and the potential for legitimate political institutions to function as solutions. Second, the analysis provides a valuable critical lens through which the actions (and omissions) of our leaders can be assessed.

Finally, since the legitimate zone of sovereign discretion can only be determined in a piecemeal fashion, assessing the extent to which the challenge of climate change falls (or ought to fall) within the legitimate zone of sovereign discretion enhances our understanding of how we ought to ascertain the extension of the legitimate zone of sovereign discretion. It is a model for how future discussions should be conducted.

The transition from theory to practice does not come in one fell swoop, rather it is developed in a rigorous and painstaking manor. The focus here is on climate change because if there is ever going to be a challenge that lies outside of the legitimate zone of sovereign discretion it is likely to be ***global*** climate change. In which case, if one cannot establish that climate change is outside of the legitimate zone of sovereign discretion, then it is likely that the legitimate zone of sovereign discretion has a far wider normative extension than many, including me, may find intuitive. In addition, climate change, due to its pervasively transnational nature, is instructional in that it is an ideal case to use to explore the application of the framework developed in *Section I.*

This transition from theory to practice unfolds in three steps over six chapters. In Chaps. 6 and 7, the discussion is focused on setting the stage for the substantive work that follows in Chaps. 8 through 11. Chapter 6 is an explication of the framework, explaining how the different elements of the framework hang together and how the framework itself is to be employed. Chapter 6 also provides a justification for the methodology chosen. Chapter 7 is a discussion of the challenge of climate change. The primary focus is on explaining why the challenge of climate change generates a practical and moral tension with sovereignty and sovereign discretion.

Having justified both the framework and its application to the challenge of climate change, Chaps. 8 and 9 are the stepwise application of the framework. Chapter 8 assesses whether the challenge of climate change falls within the limits to the legitimate zone of sovereign discretion inherent to the notion of sovereignty itself. Chapter 9, having determined in Chap. 8 that it does not, assesses whether the underlying instrumental moral justification for state sovereignty and sovereign discretion provides us with an independent reason for including the challenge of climate change within the legitimate zone of sovereign discretion.

Assuming that the challenge of climate change does not (and ought not) fall within the legitimate zone of sovereign discretion, Chaps. 10 and 11 seek to complete the transition from theory to practice by offering some suggestions for institutional design and reform. In light of the fact that, at this point in the discussion, it will have been argued that the challenge of climate change does not (and ought not) be included within the legitimate zone of sovereign discretion, how the international community ought to respond to the challenge of climate change is not fettered by sovereignty. Thus, we need an account of institutional legitimacy to guide our efforts at reform and institutional design. Chapter 10 provides just such an account.

Chapter 11 is the final step in the transition from theory to practice, concluding our discussion of the moral nature and limits of state sovereignty. To be specific, Chap. 11 employs the standard of institutional legitimacy developed in Chap. 10 to accomplish two objectives. First, it is used to critically assess the Paris Climate Agreement, which is the current institutional framework being employed by the international community to respond to the challenge of climate change. In Chap. 11, I argue that the Paris Climate Agreement fails to satisfy the account of institutional legitimacy developed in Chap. 10. This leads to the second objective of this final chapter; namely, relying on the arguments and analyses from throughout this discussion and the standard of institutional legitimacy developed in Chap. 10, I offer some thoughts and suggestions for institutional design and reform.

1.5 Conclusion

I have come to realize two things about sovereignty, one intuitive the other surprising. We should expect that sovereignty would play such an important role in the normative framework of international relations. What is surprising, despite the significance of sovereignty to the normative framework of international relations, there has been little sustained philosophical reflection on the principle itself.

The importance of this gap in the literature is only magnified by the fact that the world is growing ever more interdependent and interconnected. Yet, we continue to structure nearly all our international political relationships around a principle that originated in the seventeenth Century and was intended to address problems faced by a European community torn apart by religious wars. Whether one agrees with the conclusions drawn from this discussion, it would be hard to deny that there is a need—both practical and theoretical—to think critically about sovereignty, its moral foundations, and the role it ought to play in the normative framework of international relations.

References

Bartlelson, Jens. 1995. *A Genealogy of Sovereignty*, 5. Cambridge: Cambridge University Press.
Brock, Gillian. 2009. *Global Justice: A Cosmopolitan Account*. Oxford: Oxford University Press.

Byers, Michael. 1999. *Custom, Power and the Power of Rules*. New York: Cambridge University Press.

Chertoff, Michael. 2009. The Responsibility to Contain: Protecting Sovereignty Under International Law. *Foreign Affairs* 88 (1): 130.

Cohen, Jean L. 2012. *Globalization and Sovereignty: Rethinking Legality, Legitimacy, and Constitutionalism*. Cambridge: Cambridge University Press.

Goldsmith, Jack, and Eric Posner. 2005. *The Limits of International Law*. Oxford: Oxford University Press.

Gross, Leo. 1948. The Peace of Westphalia 1648-1948. *American Journal of International Law* 42: 20.

International Commission on Intervention and State Sovereignty. 2001. *The Responsibility to Protect*.

Jackson, Robert. 2007. *Sovereignty: The Evolution of an Idea*, 2. Cambridge: Polity.

Janis, Mark W. 1999. *An Introduction to International Law*. 3rd ed. New York: Aspen Law & Business.

Karp, David Jason. 2008. The Utopia and Reality of Sovereignty: Social Reality, Normative IR and 'Organized Hypocrisy'. *Review of International Studies* 34 (2): 313–335.

Kassner, Joshua J. 2012. *Rwanda and the Moral Obligation of Humanitarian Intervention*, 12–18. Edinburgh: Edinburgh University Press.

Krasner, Stephen. 1999. *Sovereignty: Organized Hypocrisy*. Princeton: Princeton University Press.

Kratochwil, Friedrich. 1989. *Rules, Norms and Decisions: On the Conditions of Practical and Legal Reasoning in International Relations and Domestic Affairs*, 26. Cambridge: Cambridge University Press.

Luban, David. 1980. The Romance of the Nation State. *Philosophy & Public Affairs* 9 (4): 392–397.

MacCormick, Neil. 1996. Liberalism, Nationalism and the Post-Sovereign State. *Political Studies* XLIV: 553–567.

Morgenthau, Hans. 1967. To Intervene or Not to Intervene. *Foreign Affairs* 45276 (3): 425–436.

Morris, Christopher W. 1998. *An Essay on the Modern State*. Cambridge: Cambridge University Press.

Rabkin, Jeremy. 1998. *Why Sovereignty Matters*. Washington: The AEI Press.

Raz, Joseph. 1999. *Practical Reason and Norms*, 35–42. Oxford: Oxford University Press.

Sen, Amartya. 2009. *The Idea of Justice*. Cambridge: Harvard University Press.

Shue, Henry. 1996. *Basic Rights: Subsistence, Affluence, and U.S. Foreign Policy*. 2nd ed. Princeton: Princeton University Press.

———. 2004. Limiting Sovereignty. In *Humanitarian Intervention and International Relations*, ed. Jennifer M. Welsh. Oxford: Oxford University Press.

Sinclair, Adriana. 2010. *International Relations Theory and International Law: A Critical Approach*. Cambridge: Cambridge University Press.

Singer, Peter. 2004. *One World*. New Haven: Yale University Press.

Smith, Michael J. 1999. Humanitarian Intervention: An Overview of the Moral Issues. In *Ethics & International Affairs*, ed. Joel H. Rosenthal. Washington, D.C.: Georgetown University Press.

Steger, Manfred. 2003. *Globalization: A Very Short Introduction*, 61. Oxford: Oxford University Press.

Walzer, Michael. 1985. The Moral Standing of States: A Response to Four Critics. In *A Philosophy & Public Affairs Reader: International Ethics*, ed. R. Beitz, Marshall Cohen, Thomas Scanlon, and A. John Simmons, 217–237. Princeton: Princeton University Press.

Part I
Developing a Framework for Assessing the Moral Limits of State Sovereignty and Sovereign Discretion

Chapter 2
Explicating the Tension Between State Sovereignty and the New Challenges Facing the International Community

> The Peace of Westphalia legitimated the right of sovereigns to govern their peoples free of outside interference, whether any such external claim to interfere was based on political, legal or religious principles . . . .[1]

As noted, the significance of this project is grounded in the current belief that a tension exists between sovereignty and the practical and moral demands generated by the new and emerging challenges facing the international community. As such, understanding this project requires that one understand the bases for, and nature of, this perceived tension. In addition, it is by exploring this tension and its underlying assumption about the legitimate zone of sovereign discretion that we come to have a better understanding of state sovereignty and sovereign discretion.

With that said, this chapter serves two purposes. First, it explicates the beliefs and conceptual understandings upon which the existence of the tension relies and how, taken together, these beliefs and understandings have led many to conclude that a tension exists. Second, it provides an argument that the assumption about the legitimate zone of sovereign discretion at the heart of the extant understanding of the moral nature and limits of state sovereignty and upon which the perceived tension relies is unjustified. As such, if the underlying assumption is unjustified, then we ought to be skeptical about the supposed tension that is built upon it. If correct, this opens the conceptual space for a free-standing inquiry into the moral nature and limits of state sovereignty and sovereign discretion (Chaps. 3 and 4).

---

[1]Janis (1994), p. 393.

J. J. Kassner, *Climate Change and Sovereignty*, AMINTAPHIL: The Philosophical Foundations of Law and Justice 10, https://doi.org/10.1007/978-3-030-73578-4_2

2.1 Understanding the Tension

Identifying the elements of the perceived tension begins with an explication of the tension itself. To be specific, one needs to understand the nature of the tension to be able to identify the beliefs and conceptual understandings upon which the tension relies. One might argue that it would be better to start with an identification of the beliefs and conceptual understandings underlying the tension because the tension is not an abstract possibility; rather, it is an existing phenomenon. In which case, we can discern what the elements are without a need for an abstract understanding of the tension.

Our ability, however, to identify the beliefs and conceptual understandings upon which the existence of the tension relies depends on our understanding of the tension itself. In short, the nature of the tension determines what beliefs and understandings are capable of leading to the tension itself. It is not enough to catalogue the phenomena that appear to give rise to the tension because the tension with which we are concerned is a particular sort of tension. In metaphorical terms, our understanding of the nature of the tension serves as the lens which enables us to identify the elements of the tension.

With that said, the perceived tension is a moral matter and is based in an apparent conflict between competing demands on the practical deliberations of states. On the one hand, the normative framework of international relations is based in a commitment to respect state sovereignty. On the other hand, the new and emerging challenges facing the international community generate moral and practical demands to which a response is morally and practically required.

From this we can derive several characteristics of the tension that guide our identification of the necessary elements upon which the existence of the tension relies. For one, the tension is not merely practical; it is moral. Additionally, the tension is unique to our age in two ways. First, it is the product of a globalized world; and, second, it is inexorably linked to a system of international order grounded in a system of sovereign territorial states. It is this understanding of the tension that informs our identification and understanding of the beliefs and conceptions upon which the existence of the tension relies.

2.1.1 Delineating the Conception of Sovereignty at Issue

On one side of the perceived tension is state sovereignty and the international political system founded upon it. However, as noted in Chap. 1, state sovereignty does not have a clear and univocal meaning. There are juridical, conventional, historical, and normative understandings, to name a few. Not all of these understandings can generate the tension at issue. As such, explicating the elements of the perceived tension begins with the identification of the specific conception of sovereignty upon which the tension relies.

There are two general categories into which our conceptions of sovereignty can be divided, namely, those that are descriptive and those that are normative. The former describe sovereignty as it is or has been, and the latter seek to understand the prescriptions—prudential and moral—implied by sovereignty as a foundational norm of the normative framework of international relations.

One might object that this distinction fails to capture the complex and often interrelated nature of these different conceptions of sovereignty. For example, efforts to identify and define the conventional understanding of sovereignty are projects that fall within the descriptive category. Yet, one's understanding of sovereignty as a conventional norm has implications for the prudential prescriptions generated for actors whose conduct is limited by the norm; and, so understood, falls on the normative side of the ledger.

The objection, however, misunderstands the point of the distinction and the methodological role it plays here. In the effort to identify what understanding of sovereignty generates the perceived tension at issue, there is no claim that ***the*** correct understanding of sovereignty is being identified. Rather, the understanding of sovereignty I am seeking to identify is defined by the tension. To explain, the distinctions relevant to this effort are based on what best disaggregates the properties of the conception of sovereignty capable of generating the tension at issue.

The first distinction to be drawn is between descriptive and normative conceptions of sovereignty. Included among such descriptive understandings are both historical and conventional accounts. Neither of which can generate the tension at issue because the perceived tension only arises when the underlying conceptions and beliefs generate a ***moral*** conflict. Descriptive understandings don't give rise to moral claims; consequently, descriptive understandings can't generate the necessary moral reasons for action upon which the perceived tension is based.

Taking F. H. Hinsley's seminal work on sovereignty as an example of such descriptive accounts, for Hinsley, attributions of sovereignty correspond to a particular form of political structure.[2] Namely, a hierarchical one where there exists a supreme political power governing a community.[3] Nothing in this descriptive understanding is relevant to what states ought or ought not do, it is simply a statement of observable facts—historical in Hinsley's case—about the structure of governance within a political community.

The normative understanding is about the legal or moral standing of states within the international arena. Under this normative understanding the practical deliberations of states and the international community are guided and limited by sovereignty. It is this understanding of sovereignty that serves as the normative foundation for the state system itself, providing a constraint on what is and is not acceptable within international law and relations.

The tension is thought to arise because of a normative conflict in the practical deliberations of agents whose conduct is governed by sovereignty, namely, states .

---

[2]Hinsley (1986).

[3]*Ibid.*

To be more specific, the relevant notion of sovereignty provides states with discretion to decide for themselves whether and how to respond to the practical and/or moral challenges they face. It also implies a prohibition on interference by others in those matters that fall within the legitimate zone of sovereign discretion.

The new and emerging challenges facing the international community generate moral and practical imperatives that states ***must*** act to resolve such matters. Descriptive understandings of sovereignty, like that offered by Hinsley, are not tied to such prescriptive claims. The normative understanding, however, is. Thus, the perceived tension depends on a normative understanding of sovereignty.

Next, to avoid confusion over different normative conceptions of sovereignty, one should recognize that, as Christopher Morris has noted, "[i]t is common to distinguish internal from external aspects of sovereignty".[4] This distinction between internal and external sovereignty tracks different normative understandings of sovereignty. As Robert Keohane notes, "[t]he classic unitary conception of sovereignty is the doctrine that sovereign states exercise both internal supremacy over all other authorities within a given territory, and external independence of outside authorities."[5]

Understanding the respective normative relationships constitutive of internal versus external sovereignty is key to understanding the distinction between the two, and the basis for the relatively uncontroversial proposition that it is external sovereignty that generates the tension. As Roger Scruton points out, domestic political relationships between the institutions of the state and the state's inhabitants are what make up internal sovereignty.[6] Thus, the impact, at least under present practice, of moral reasons for action generated by internal sovereignty have little to do with the practical deliberations of those in the international community. External sovereignty, on the other hand, is "an attribute which political bodies possess in relation to other such bodies."[7] Thus, the relevant parties to the type of relationship that defines external sovereignty are sovereign states themselves, and the rights attendant to the possession of external sovereignty are rights held against other sovereign states and the international community.

Since the perceived tension arises when sovereignty stands as an obstacle to the international community's ability to meet the moral and practical demands generated by new and emerging challenges facing the international community, that tension is linked to the impact state sovereignty has on the practical deliberations of states and the international community. As noted, internal sovereignty does not generate reason for action for the international community, but external sovereignty does. As such, our concern here is with external sovereignty.

One may be skeptical of the significance of this distinction, believing that there is little that depends on recognizing that internal and external sovereignty are different.

---

[4]Morris (1998), p. 172.

[5]Keohane (2003), p. 282.

[6]Scruton (1982), p. 441.

[7]*Ibid.*

Within political philosophy, however, internal and external sovereignty are generally assessed by different criteria. Claims to internal sovereignty are evaluated based on accounts of political legitimacy focused on the vertical relationship between ruler and ruled.[8] A government's claim to authority over its inhabitants is only justified if that relationship bears the marks of legitimacy. The academic debate is over what those "marks" are.

External sovereignty is, on the other hand, about the horizontal relationship amongst equals, at least equals from the perspective of the normative framework of international relations. Standards of legitimacy for external sovereignty have tended to ignore the internal relationships constitutive of internal sovereignty.[9] Consequently, most accounts of the legitimacy of external sovereignty do not depend on standards of internal legitimacy.

This is not to say that there aren't those who challenge the validity of the internal-external divide. For example, Fernando Teson has offered a compelling account of the legitimacy of international law in which the legitimacy of a state's claim to external sovereignty depends on the legitimacy of its claim to internal sovereignty.[10] I am sympathetic to Teson's view and the larger project that seeks to bring these two aspects of sovereignty together. That, however, is not the project here. Nonetheless, one possible benefit of this discussion of the moral nature and limits of state sovereignty is the light that it may shed on the relationship between internal and external sovereignty.

Focusing on the fact that there are multiple normative understandings of state sovereignty, the third aspect of the notion of sovereignty at issue is tied to the fact that sovereignty is a conventional norm, but not morally a neutral one. As such, it can be understood from at least two different perspectives. One might seek to understand the role sovereignty ***is taken to play*** in the normative framework of international relations and nothing more. This is an empirical matter. On the other hand, since sovereignty is often invoked as a moral norm, one might seek to understand its moral nature and basis, a moral matter.

With these possible understandings in mind, there are two differences especially relevant to this project. First, there is a category difference to be noted; and, second, there is a difference related to the function each understanding serves in the normative framework of international relations. As to the category difference, state sovereignty understood empirically is a social fact defined by the role it ***in fact*** plays in the international arena. Under such an understanding, it is understood through empirical inquiry and has no necessary connection to morality. On the other hand, sovereignty understood as a moral norm is about the moral standing of states (as rights-holders

[8]For example, see Hobbes (1996). See also Locke (2015). And see Simmons and Wellman (2005).

[9]This is not to say that this is the correct way to understand the relationship between internal and external sovereignty, but the two have tended to be treated in categorically different ways. As an example of the alternative view, Fernando Teson offers an account of the legitimacy of external sovereignty that is dependent on the legitimacy of internal sovereignty. Teson (1998).

[10]Teson (1998).

and duty-bearers) in the international arena. It is about the range of actions a sovereign state can claim are within its moral discretion and about the treatment that sovereign states are entitled to demand of one another. This understanding is dependent on its underlying moral justification.

Before proceeding, there is a possible misunderstanding of the difference just highlighted that warrants attention. One might find the difference to be reminiscent of the distinction underlying the disagreement between legal positivists and classic natural lawyers. In responding to the question of what makes a norm a legally valid norm, legal positivists contend that there is no necessary connection between law and morality.[11] Classic natural lawyers argue, on the other hand, that legality is dependent on morality; namely, that a norm cannot be a legally valid norm unless it is (at least) consistent with morality.[12]

Though an oversimplification of that debate, one can adduce that the underlying disagreement revolves around a single inquiry. Namely, when seeking to understand the nature of law, is there a necessary connection between legality and morality? The answers of the legal positivist and the classic natural lawyer are mutually exclusive. In this simplified version of that debate, no matter which position one defends it will prove antithetical to those who defend the other.

Unlike that debate, however, the difference between an account explaining state sovereignty as a social fact and one addressing sovereignty as a moral norm does not involve an inherent conflict because they need not be thought of as competing answers to the same question. Those seeking to understand state sovereignty as a social fact are answering the question, "What is the best understanding of the role sovereignty ***in fact*** plays in the normative framework of international relations?" On the other hand, those seeking to understand the moral nature of sovereignty are addressing the question, "What, if anything, can ***morally justify*** sovereignty and the role it plays in the normative framework of international relations?" How one answers one question does not commit one to any particular answer to the other.

Another difference between the empirical and moral understandings of sovereignty is the different function either serves in the normative framework of international relations. As to our understanding of sovereignty as a social fact, it has rhetorical, predictive, and prudentially prescriptive power. Rhetorical because an accurate understanding of sovereignty as a social fact provides other actors in the international arena with the ability to recognize when their invocation of the principle will be accepted by others—which claims can be expected to be viewed as reasonable and which not. As to its predictive power, it allows others to have reasonable expectations about the actions of other states. This will give rise to prudential prescriptions enabling states to pursue their interests within a relatively stable environment. A stable system does not, however, imply a just or legitimate one.

[11]For the defense of this position see Hart (1994).

[12]For a contemporary defense of this position see Finnis (2011).

Sovereignty understood as a moral norm serves several functions distinct from those connected to the empirical understanding. For one, sovereignty is often intended, as a moral justification for state action (or inaction). As such, when appealed to in this way its persuasive force is tied, not merely to its rhetorical strength, but to the strength of the underlying justification. In addition, the rights and duties implicit in the possession of sovereignty generate moral entitlements and obligations. Consequently, understood as a matter of morality, sovereignty produces moral, rather than prudential, prescriptions. Ultimately then, sovereignty as a moral norm has justificatory and morally prescriptive power, rather than rhetorical, predictive, and prudential power.

Returning to the nature of the tension at issue for guidance, the tension is based on the proposition that what sovereignty morally permits and prohibits is in conflict with the moral and practical imperatives generated by the new and emerging challenges facing the international community. Thus, the tension is a matter of moral, not merely practical, conflict. As such, understanding the tension is dependent on understanding sovereignty as a moral principle.

One might argue that the perceived tension could just as easily arise as an empirical matter. For example, one could claim that the tension just is a fact of the circumstances of international relations in which our conventional understanding of sovereignty is inconsistent with the perceptions regarding the moral and practical demands generated by the new and emerging challenges facing the international community.

Such objections, however, fail to appreciate the practical issue underlying the tension. When a state or the international community is faced with the question of what ***ought*** to be done when facing a choice between respecting sovereignty and acting on a moral imperative that would (appear to) require the violation of another state's sovereignty, pointing out that there is, as a matter of fact, a conflict does little to aid in one's understanding or resolution of the tension. So, even in this instance, one is directed towards questions of moral justification.

2.1.2 Hohfeld, Sumner and the Constitutive Elements of State Sovereignty

Notwithstanding the distinctions already noted, the notion of sovereignty thus far identified is insufficient. It lacks the substantive understanding necessary for appreciating why this conception of sovereignty generates the perceived tension. To provide further clarity, I am adopting an approach grounded in the typology developed by Wesley Newcomb Hohfeld.[13] Hohfeldian analysis has several distinct advantages. First, both legal theorists and political philosophers are familiar with Hohfeld's typology, thus employing this methodology allows for an analysis that is

[13]Hohfeld (1978).

familiar across disciplines. Consequently, in addition to avoiding confusion, this approach is an effective conceptual bridge between different disciplinary approaches, enabling more fruitful interdisciplinary discussions.

For Hohfeld, understanding a norm requires that one understand the normative relationship constitutive of the norm.[14] Hohfeld thought of such relationships as being defined by jural correlatives.[15] For example, to understand a right it is not enough to identify the substance of the right and the identity of right-holders. One must also identify the correlative duties and duty-bearers. Hohfeld's work was significant, not simply because it provided a clear framework for understanding the nature of legal rights' claims, but also because it shed light on the complexity of such norms.

In his efforts to understand the nature of moral rights, L. Wayne Sumner developed an approach that relies on Hohfeld's typology.[16] Relying on the Hohfeldian commitment to understanding norms as being comprised of correlative relationships, Sumner proposed a methodology for assessing the moral foundations of rights that begins with the identification of the constitutive elements of the right, namely, a right's content and scope.[17] In addition to enabling us to understand the right in question, Sumner's approach provides an effective means of assessing possible justifications. Any proffered justification must accommodate the right's constitutive elements. If it cannot then one ought to reject the proffered justification or revise their understanding of the right.

As to the constitutive elements of a norm—its content and scope—the content of a norm is the substantive moral claim definitive of the normative relationship at issue. This is not likely to be a simple moral claim—***a*** claim right or ***a*** liberty or ***a*** power. Rather, the content will likely be a conjunction of various moral claims and correlative burdens. For example, consider the right of individuals to freedom of expression. The content of the right includes, at the very least, the liberty to express one's views. This, however, is not enough, the content also includes a claim to protection and an immunity from interference.

In addition, the right to freedom of expression creates correlative moral burdens. Therefore, understanding the content of the right requires an understanding of the correlative burdens implied by the right. Regarding the right to free expression, that includes, at the very least, a duty to refrain from interfering in the exercise of the right and a duty to protect the liberty a right-holder has to express their views free from interference by others.

As to the scope of a right, it is comprised of the agents whose normative relationship is defined by the right. Returning to our example, the right is held by individuals and (arguably) may be held by other agents.[18] The correlative duty to

[14]Hohfeld (1978), pp. 35–39.

[15]*Ibid.*

[16]Wayne Sumner (1986).

[17]*Ibid.*

[18]*Citizens United v. Federal Election Comm'n*, 558 U.S. 310 (2010).

refrain from interfering is borne by states and other individuals, and the duty to protect the right-holders' liberty is borne by the state. This example is underdeveloped; however, it should give the reader a sense of what Sumner's framework requires.

2.1.2.1 The Content of Sovereignty

The explication here begins by identifying sovereignty's content since, in my considered judgment, once the content has been identified the scope is more easily delineated. That said, sovereignty is a complex norm comprised of a bundle of rights; namely, the rights to territorial integrity, political independence, and sovereign equality. Consequently, understanding the content of sovereignty requires that one understand the content of this bundle of rights.

Before proceeding, there is a distinction between sovereign equality and the rights to territorial integrity and political independence that should be recognized. Under the right to sovereign equality, sovereign states have equal standing within the normative framework of international relations. Unlike the rights to political independence and territorial integrity which are defined by the specific substantive entitlements sovereign states are owed, sovereign equality is about the claim that each sovereign state has to be treated equally, regardless of the content of the substantive rights attributed to sovereign states.

One might object to this distinction, arguing that the right to sovereign equality possesses substantive content; namely, sovereign states have a justified claim to ***equal treatment***. One cannot deny that equal standing involves important moral values, equality and fairness, for example. Without more, however, the practical meaning of equal treatment is hollow. The fact that all states are entitled to equal treatment tells us how they are to be treated relative to one another, but it begs the question about the specific treatment to which all sovereign states have an equal claim. In the case of state sovereignty, the rights to territorial integrity and political independence provide this substantive element.

As such, in explicating the content of sovereignty, my focus will be on the rights to territorial integrity and political independence. Appreciating the contribution such rights make to our understanding of the content of sovereignty requires that we begin with the identification of the specific entitlements to which rights-holders have a claim and the moral burdens correlative to such rights. As to the former, the rights to territorial integrity and political independence have both internal and external elements. The focus here is primarily on the external elements.

As to the right to territorial integrity, there are at least two aspects of the content of the right relevant to this inquiry. First, the right includes an entitlement to control the international political borders that define the territorial extension of the state. Second, the right includes an entitlement to control the natural resources found within the international political borders of the state—the oft-discussed resource privilege.[19]

The entitlement to control political borders is comprised of, at least, two elements, a protected liberty over the placement of the political borders that define the state and an entitlement to determine who or what can cross such borders and for what reasons. As to the former, it is important to distinguish the internal aspect of this liberty from the external. The right includes an entitlement to establish internal administrative borders as the right-holder sees fit.[20] As for the political borders that define the territorial extension of the state within the international arena, the right operates as an immunity against alteration to these borders by others. Such borders can only be changed with the consent of the right-holders. As to the latter element, it is only with the permission of the state that agents, objects, or (arguably) even information can cross such borders.

As to the second aspect of the right to territorial integrity, the entitlement to control the natural resources found within the political borders defining the state, it includes an entitlement to regulate and dispose of the natural resources found within the territory of the state as the right-holder sees fit.[21] This includes normative powers that extend over the resources themselves as well as over individuals and groups who may have an interest in the resource(s) at issue, including a power to exploit the right-holder's natural resources for its own purposes, a power to sell the resources or access to them to others, and a power to exclude those who haven't been granted access.

As to duties correlative to the right to territorial integrity, they are equally complicated. There are duties related to the right of a states to control its borders, and duties related to the right of a state to control its natural resources. With regards to the former, duty-bearers have a duty not to interfere with the location of a right-holder's borders. In addition, duty-bearers are under a duty not to cross or transport goods across a state's borders without the right-holding state's permission. It is also a liability under which the right-holder can alter the normative status of the duty-bearer; wherein, the right-holder can grant the duty-bearer permission to alter the political boundaries of the right-holding state and/or to cross such borders. In short, what was impermissible becomes permissible, even if in some cases only conditionally.

As to the resource privilege, there are two correlative duties relevant to this discussion. For one, duty-bearers are under a duty not to interfere with a right-holder's disposition and management of its natural resources. In addition, duty-bearers are under a duty to refrain from using the natural resources of the right-holding state without the latter's permission. Again, the duty not to interfere is coupled with a liability borne by the duty-bearer to have its normative status altered

[19]Pogge (2008), pp. 29–30.
[20]Kassner (2010), pp. 2–7.
[21]Pogge (2008).

by the right-holder. A duty-bearer's access to a right-holder's natural resources depends on permission being granted by the right-holder.

With this understanding of the content of the right to territorial integrity, our attention now turns to the right to political independence. The right to political independence is primarily about self-governance, about political self-determination. To understand the content of the right to political independence, one must recognize that there are two distinct ways in which a sovereign state is entitled to be self-governing. The right to political independence is both about how the state is to be governed as a matter of institutional structure, and about the policy choices made and actions taken by the state.

As to the former, the right to political independence includes an entitlement to choose the structure of the political institutions and decision-making procedures employed by the state. For example, whether government and political power is unified or divided, constrained by a written constitution or not, an independent judiciary or one that is elected and whether the population has a right to participate in the political process are matters within this aspect of the content of the right.

This should not be taken to imply that there aren't morally significant differences between these alternatives. For example, decentralized secular democratic decision-making procedures constrained by a written constitution may, in fact, be morally preferable to the alternatives; however, in holding the right to political independence, a sovereign state is entitled, within limits that will be discussed below, to make such determinations for itself, even if the path chosen suffers from moral and/or practical failings.

The right to political independence also includes an entitlement to make policy choices. Like other aspects of the rights of sovereign states, this aspect of the content of the right to political independence is comprised of both internal and external elements. As to the former, the state is largely free to pursue domestic law and policy as it sees fit, immune from outside interference with its choices. There is, however, growing consensus around the proposition that human rights limit a state's internal discretion, as captured by ICISS's development of the Responsibility to Protect doctrine.

Regarding the external aspect of the right to political independence, it includes a protected liberty to pursue, within limits, a foreign policy agenda of a right-holder's own choosing. Perhaps most important for this discussion is that in their foreign policy initiatives right-holding states are immune from compulsion by foreign powers. On a wide range of matters, their action or inaction is a matter of discretion.

Finally, as to the moral burdens correlative to the right to political independence, two are relevant to our efforts to understand the tension at issue. First, there is a duty to refrain from interfering with the protected choices of rights-holders, including their choice of political structure as well as their policy choices. Second, there is a liability borne by duty-bearers, wherein their normative relationship can be altered by a right-holder. For example, a duty-bearer's diplomatic relationship with a right-holder is dependent on choices made and permissions granted by the right-holder.

The identification of the content of the right to political independence may shed light on, but is not, however, equivalent to, its justification. For example, the fact that the content is largely about self-governance does not mean that self-governance is what justifies the right. In fact, as I will argue, it is my contention that the right to political independence, as a historical artefact, was intended to promote and protect international peace and security. This does not preclude additional justificatory bases, like the interest a political community has in political self-determination, but the content is not necessarily the justification.

More could be said about the content of the bundle of rights that make up state sovereignty. Such an understanding is not, however, necessary for one to understand the perceived tension being explored here, nor is it necessary for the arguments and analyses that follow.

2.1.2.2 Reconstructing the Content: The Right of Nonintervention

Before delineating the scope of sovereignty, a reconstruction of the content is in order. The combined content of the rights to territorial integrity, political independence and sovereign equality imply a right held equally by all sovereign states to be free from intervention – the right of nonintervention. Correlative to this right is a duty not to intervene in those choices and/or actions of another sovereign state that are protected by that state's sovereign status.

This understanding of the content of the right of nonintervention raises an important question. Namely, how do we determine which choices and/or actions are protected by the right and its correlative duties? This question is at the heart of the broader inquiry with which this book is concerned. The account of the moral nature and limits of state sovereignty presented in Chaps. 3 and 4 provides the answer to this question.

One might question why, after offering such a detailed exposition of the content of the constitutive rights, the discussion would then shift to a less detailed understanding for the purpose of delineating the scope of those rights. The choice is a methodological one. First, the reconstruction ensures that the arguments and analyses that follow in the remainder of this chapter are readily understood, and not made unnecessarily confusing by the complexity of the rights and duties that make up the content of state sovereignty. Additionally, it is this simplified understanding of the content that plays such a significant role in the political rhetoric of international relations.

With this in mind, one might raise a different objection. Namely, if the argument can be made with the right of nonintervention, then why provide such a detailed understanding of the content of the rights to political independence and territorial integrity in the first place? For one, substituting the right of nonintervention for the constitutive rights discussed above serves several valuable methodological purposes. Focusing on the right of nonintervention avoids unnecessary confusion by illuminating the most relevant aspects of the scope of state sovereignty and the role sovereignty plays in the practical deliberations of states. In addition, the detailed understandings of the content of the rights to political independence and territorial integrity were not provided in vain as they are relied upon in Chaps. 3 and 4. As such, the detail provided, though being set aside for methodological reasons at the moment, is necessary to fully understand the moral nature and limits of state sovereignty.

2.1.2.3 Delineating the Scope of State Sovereignty

With this said, regarding the scope of state sovereignty, under Sumner's framework the scope of a norm is comprised of the individuals or agents whose normative relationship is defined by the norm in question.[22] As to the right of nonintervention, states hold the right and bear, both individually and collectively, its correlative duties. The most significant aspect of this understanding of the scope of the right of nonintervention, is that pursuant to this understanding states, either individually or collectively, are the only relevant parties to the relationship. I am not endorsing this view, nor does it exhaust our understanding of the moral burdens that states may bear within the international arena, but it is important to note because any extension of the duties that states bear will be dependent on free-standing arguments, and not necessarily implied by sovereignty itself.

2.2 The Role of Sovereignty in the Practical Deliberations of States

Sumner's analysis of the constitutive elements of a right is limited to the identification of its content and scope. There is, however, an additional element that is especially important considering that the objective is to understand the moral nature of sovereignty and its role in the tension being explored here.[23] To clarify, sovereignty is a norm and is taken to influence the behavior of those agents to whom it applies. It guides by providing states with reasons for action through moral prescriptions. Those reasons are to be considered within the practical deliberations of states as they consider what choices to make and actions to take within the international arena. As such, a sufficient understanding of sovereignty as a moral norm requires an appreciation for the impact sovereignty has on the practical deliberations of states.

Another reason for exploring the role sovereignty plays in the practical deliberations of states is tied to the nature of the tension underlying this inquiry. It is a practical tension tied to the various permissions and prescriptions at play. Consequently, understanding the practical tension requires that one understand the role sovereignty plays in the practical deliberations of states.

---

[22]Wayne Sumner (1986).

[23]Kassner (2012).

Finally, in Chap. 4 the elements constitutive of sovereignty are used to assess the possible moral justifications for sovereignty. To leave out the role sovereignty plays in the practical deliberations of states and in the normative framework of international relations when delineating those elements would not only lead to a less than satisfactory understanding of sovereignty but would also leave out the most significant aspect of sovereignty requiring justification.

It is important to understand the specific account of practical deliberation upon which the arguments and analyses that follow rely. Yet, to this point, little has been said about practical deliberation. The classic account invokes the metaphor of balancing and is familiar to most. What one ought to do when faced with a choice of action is determined by the ***balance*** of reasons.[24] Under this account, the relevant practical reasons are all first-order. They are balanced directly against one another. The choice backed by the "weight of reasons" is the one that ought to be pursued.

This is ***not*** the account of practical deliberation being relied upon here.[25] Rather, for reasons to be discussed momentarily, the account of practical deliberation developed by Joseph Raz ought to be preferred to the classic account.[26] The most important distinction between the classic account and the Razian account is that Raz contends that there are, in addition to first-order reasons, second-order reasons.[27] Consequently, when engaged in practical deliberation an agent does more than balance the competing first-order reasons, rather the deliberating agent must also take second-order reasons into consideration.

Recognizing that there are second-order reasons, however, raises several important questions. Under what circumstances are such higher-order reasons generated, and what impact do such reasons have on the practical deliberations of the agents to whom they apply? An exhaustive answer is not possible here. There is, however, one category of moral reasons relevant to this inquiry that constitute a particular type of second-order reason, namely non-consequentialist moral obligations. Some examples include individual rights, promissory obligations, and special relationships (i.e. parent/child). What each of these have in common is that they alter the simple first-order weighing calculus. They have an exclusionary impact on an agent's practical deliberations.

One might further inquire about the underlying moral basis for treating such moral norms as second-order reasons. This is certainly a reasonable inquiry; however, the discussion that follows is not dependent on the answer. Rather, the perceived tension that lies at the heart of this inquiry and the arguments and analyses that follow are based on beliefs about the role sovereignty plays in the practical deliberations of states. Thus, as a matter of social fact, when invoked, sovereignty is treated as a second-order exclusionary reason. It is taken to exclude certain first-order reasons for action from consideration by states.

---

[24]Raz (1999), p. 36.

[25]Kassner (2012).

[26]*Ibid.*

[27]Raz (1999), pp. 35–39.

To clarify, when implicated, second-order reasons change the calculus used to determine what an agent ought to do all-things-considered. The metaphor of balancing no longer accurately captures the deliberative process. Second-order reasons act as both weighty first-order reasons in favor of some action, but also exclude certain countervailing first-order reasons from consideration. In other words, first-order reasons that would, *ceteris paribus*, normally be relevant are rendered irrelevant—excluded from consideration—because they are within the exclusionary extension of the second-order reason. As such, prior to "balancing", a second-order reason knocks some first-order reasons off the scales.

An example may offer further clarification. Think of an individual's legal right to freedom of expression as found in the Constitution of the United States.[28] The impact of the right is most apparent when the expression of a right-holder is offensive or contrary to the public interest. Consider the march of Neo-Nazis in Skokie, Illinois in the mid 1970s.[29] The "speech" of the National Socialist Party was not only going to be highly offensive to many in Skokie, which included Holocaust survivors, but was arguably going to have the negative effect of converting some to their cause, or causing some to feel justified in, or less ashamed by, their anti-Semitic beliefs.[30]

Without the right, one could reasonably argue that the balance of first-order reasons in this situation weighed in favor of banning the march and the speech, the costs of allowing such ignorant and hateful speech outweighing the benefits. The right of the marchers to free expression, however, rendered such considerations irrelevant, not merely outweighed. Consequently, when the case was adjudicated, it was not simply a matter of balancing reasons for and against allowing the Neo-Nazis to march; rather, it was a case of balancing the ***non-excluded***-reasons-for versus the ***non-excluded***-reasons-against allowing the march.

With the foundation laid, I can now turn to an explication of the role sovereignty—through the right of nonintervention—plays in the practical deliberations of states. The right of nonintervention, as a second-order exclusionary reason, is a reason in favor of allowing a state to govern itself and its territory free from outside interference and excludes duty-bearing states from considering certain first-order reasons from consideration, namely those reasons that would require violations of a sovereign state's rights are excluded from a duty-bearer's practical deliberations.

For example, China, as a right-holder, is entitled to choose for itself how its society will be organized politically and economically. Consequently, China's choice to pursue a one-party communist structure is well-within its discretion, regardless of the fact that many in the international community think, and are likely correct, that the one-party communist system is violative of certain fundamental freedoms. Correlative to China's right is the duty borne by other states and the

[28]U.S. Const. amend. I.

[29]*National Socialist Party of America v. Village of Skokie*, 432 U.S. 43 (1977).

[30]*Ibid.*

international community to exclude from their practical deliberations reasons for action that would require interference with China's protected choices.

This is not to say that sovereignty excludes all reasons for action based on a critical moral assessment of another state's choices. Rather, a reason is excluded from consideration only if it requires interference in the protected choices of a right-holding sovereign state. The exclusionary effect of sovereignty does not exclude reasons for action based on critical moral assessments not requiring such interference. Returning to the example of China. There are numerous reasons for action based in a moral assessment of China's choices not excluded from consideration. For example, Chinese sovereignty does not exclude, from consideration by the United States, reasons for action that direct the United States to alter its economic relationship with China. The United States could also consider reasons for action that direct it to alter its diplomatic relationship with China. In the end, such reasons may not prove decisive, but they are not excluded from consideration.

It is also important to note that the reasons being considered by states are not conclusive reasons for action. The mere fact that a reason is considered does not mean that a state is going to, or ought all-things-considered to, act on it. It may, when combined with other non-excluded reasons for action, result in an all-things-considered judgment that the course of action it supports ought to be chosen. On the other hand, it may be outweighed by non-excluded countervailing reasons; in which case, the choice of action it supports is defeated, since reasons for action are, under normal circumstances, just one among many reasons to be weighed.

There are also limits to the extension of sovereignty's exclusionary impact. The most widely accepted is that sovereignty does not exclude reasons related to international peace and security. And, more recently, there has been a growing consensus that massive human rights violations are not within the extension of the exclusionary impact of sovereignty.

In the end, sovereignty provides states with a zone of moral discretion in which a state is entitled to act without the threat of interference. This zone of discretion is determinative of the impact sovereignty has on the practical deliberations of states. Provided a state's choice of action is within a sovereign state's legitimate zone of discretion, the state is entitled to be free from outside interference.[31]

2.3 Changing Times and Global Challenges

Also, essential to understanding the tension at the heart of this discussion are the dramatic changes to the international arena since Westphalia. Whether it is the communications revolution allowing us to know what is happening across the globe at a moment's notice, or the fact that the cup ***and*** the coffee I enjoy in it come from two separate continents, neither of which is my own, the world has become ever

[31] See Kassner (2009), pp. 181–193.

more economically, culturally, militarily, epistemically, technologically, environmentally, politically and morally interconnected and interdependent. This interdependence has generated new moral and practical challenges which the international community must face.

A few examples may help to convey the significance of this transformation:

1. One contrast is to be found in the very concerns that the Peace of Westphalia was intended to address, namely, the desire for international peace and security, and the commitment to political self-determination. The threat to international peace and security no longer issues solely from other states. Now, such threats also come from non-state actors, or as the result of internal civil and political strife, from individuals and groups seeking political self-determination. Take, for example, the movement that collectively has come to be known as the "Arab Spring". Beginning in December 2010 with a single act of political protest in Tunisia, the protests would spread across the Arab peninsula and engulf many states in violent insurgencies and civil war. This is not to pass moral judgment on those seeking political freedom and empowerment. These events are destabilizing for domestic governments and a threat to international order.
2. Another difference is the impact globalization has had on the economic order. Since Westphalia, the developed world has exploited its access to natural resources, technology, and labor to industrialize, and much of the developing world is currently pursuing that same path to economic development and progress. In addition, the economic order has globalized in ways that may have been unimaginable just a few short centuries ago. One need only look at the phone in their hand, the food on their plate, the shirt on their back, or the car in their driveway to understand the ways in which the economy is global on both a macro and micro level. And, it should come as no surprise that the globalization of markets lifted many out of poverty, but has also contributed to the widening gap globally between rich and poor.[32] Such disparities place into question the various international agreements and institutions intended to regulate global trade, such as the World Trade Organization (WTO) and the Agreement on Trade Related Aspects of Intellectual Property Rights (TRIPS). As such, this interdependence is not merely a matter of economics; it is also an ethical concern.
3. Another significant transformation in the international arena is the emergence of a consensus on and effort to institutionalize human rights. Human rights discourse and practice has assumed a relevance (at least rhetorically) unique to the modern era. As noted, in the aftermath of the Second World War the international community explicitly recognized the moral importance and practical relevance of human rights. The language of the UDHR is now routinely found in both scholarly work and public discourse. Human rights have become accepted as a constraint on the discretion of states and the permissibility of state action. Despite

[32]Steger (2003), pp. 112–113.

the efforts of ICISS and its development of the RtoP doctrine, the question of what this means for state sovereignty remains undetermined.

4. For much of human history geographic distance, economics and politics have kept individuals and cultures apart. Only a select few interacted directly with individuals from, or exposed to, foreign cultures. However, advancement in communications technology and economic globalization have allowed individuals in distant lands to be connected personally, culturally, economically, and politically. News about the Arab Spring initially reached the rest of the world through social media. Individuals caught in the midst of upheaval and violence were posting videos and comments on a variety of social media platforms, allowing others across the globe to witness the events as they unfolded. There are also many global non-governmental organizations (NGOs) and secondary associations founded upon shared ideas and values. This greater connectivity between individuals, peoples and cultures raises questions about some of the assumptions upon which Westphalian sovereignty and the state system are based.[33]
5. Finally, the negative impact of industrialization on the non-human world can no longer be denied and will be felt for generations. In a pre-industrial world, the degradation of the local environment by one political community would likely have little impact on neighboring communities. In our post-industrialized world, harm to the environment is not ***only*** a local concern. In fact, some of the worst environmental effects of a political community's activities are felt by those inhabiting distant corners of the globe. Environmental harms are negative externalities that do not respect political boundaries. Looking to global climate change as an example, the activities of the world's developed nations have resulted in a dramatic change in hydrologic cycles causing desertification in Africa and rising sea levels threatening the impending disappearance of island nations like Tuvalu.

These changed circumstances have significantly altered what counts as a matter of concern for the international community. In 1648 the most pressing concern was to be protected from foreign threats to the security of individual political communities. In the late eighteenth century, one could add the political self-determination of unified political communities as a value served by sovereignty. Today, though not an exhaustive list, some of the most pressing issues are the threat of global terrorism, the provision and protection of universal human rights, immigration, political self-determination and secession, and the challenge of global climate change.

It is practically impossible for me to address each of these challenges here. As such, I have chosen to focus on a single challenge facing the international community as a way of exploring how we ought to go about assessing the moral nature and limits of state sovereignty and what ought to be done institutionally in light of those limits. It is the fact that the challenge of climate change is so globally pervasive that has led me to focus on it in *SECTION II*.

[33]Beitz (1999).

2.4 Identifying the Assumption

Conceding the transformative nature of such changes, one might remain skeptical of the claim that there exists a tension between sovereignty and the moral and practical demands generated by these new and emerging challenges. There remains a conceptual gap between the existence of such challenges and the apparent tension with sovereignty. The existence of the tension relies on two distinct conditions, one related to the nature of the challenges being faced by the international community and the other tied to our understanding of sovereignty. The latter has been, at least provisionally, addressed. As to the former, one might reasonably ask what it is about these challenges that make them particularly problematic? The simple answer is that the challenges being faced are ***global***.

Whether it is the impact of industrialization on the environment or a commitment to the idea that each individual has rights based on some aspect of their shared humanity, the causes or underlying conceptual bases of these challenges are global. The territorial boundaries that normally separate one political community from another are not effective at defining the appropriate political or moral space (human rights), constraining the effects of the various activities engaged in by a political community (economic globalization, social media, etc...), or being useful in our efforts to understand the causal story behind a challenge (climate change). In short, the global nature of these challenges, mean that they cannot be adequately understood by a political framework that views the relevant political and moral space through the lens of territorial defined states.

Additionally, the solutions are global in nature. In some cases, this involves a commitment to universal values, but in others the solutions are dependent on highly coordinated transnational action by the international community. In other words, to meet the moral and practical demands generated by these challenges, sovereign states would be ***required*** to act (or refrain from acting) in ***particular*** ways, leaving little room for discretion. A proposition in conflict with sovereign discretion.

In addition, regarding the understanding of sovereignty at issue, there is a widely shared assumption upon which the perceived tension depends. Simply put, it is assumed that the new and emerging challenges facing the international community fall within the legitimate zone of sovereign discretion. Under this assumption, each state has a legitimate moral claim to largely unfettered moral discretion over their response to these challenges, and the tension arises because, on the one hand, you have global challenges that allow for no discretion if they are to be responded to in a morally justified and effective manner; and, on the other hand, you have states with a claim to have moral discretion over such matters.

2.5 Conclusion: Rejecting the Assumption and Opening the Conceptual Space

Since the tension depends on this assumption regarding the legitimate zone of moral discretion, much depends on whether the assumption is justified. There are, at least, two compelling reasons for rejecting this assumption. First, any assumption that is not self-evident requires a justification. The assumption with which we are presently concerned is not self-evident, nor has there been a justification offered for it. Thus, begging the question about its justification.

It is an assumption about the moral nature and limits of state sovereignty. As such, whether this assumption should be accepted depends on our understanding of the moral nature and limits of state sovereignty, thus begging the underlying question with which I am concerned. Consequently, at least until this inquiry is completed, the assumption should be set aside.

Another reason for rejecting the assumption begins with an understanding of the assumption that leads either to the first reason for rejecting it or to an understanding of sovereign discretion that ought to be rejected. One can understand the assumption to be advancing a limited or an unlimited understanding of the legitimate zone of sovereign discretion. If the former, then we must ask what the nature and bases for the limitations are. In which case, we are again faced with the question of the moral nature and limits of state sovereignty and led back to the question underlying this inquiry.

If the latter, we are presented with an understanding of sovereignty and sovereign discretion that doesn't conform to existing conventional practice and leads to counterintuitive—if not absurd—implications. Unlimited sovereign discretion would imply that, unlike nearly all other moral principles, sovereignty is absolute. In addition, the very rights that make up sovereignty would become unworkable as the basis for international relations, because an unlimited understanding of sovereign discretion would not be compatible with a normative political framework based on a system of claim rights held equally by all members of the community. Rather, if sovereignty were absolute, it would generate a normative political system grounded in licenses and unprotected liberties. In the end, these seem to be compelling reasons, at least for the moment, to set the assumption upon which the tension depends aside.

What does this rejection of the assumption entail? One might infer that none of the new and emerging challenges facing the international community fall within the legitimate zone of sovereign discretion. Some may even tie the rejection to the stronger claim that sovereign discretion cannot be justified at all. Neither, however, is implied. The fact that we are unjustified in ***assuming*** that the new and emerging challenges facing the international community fall within the legitimate zone of sovereign discretion, does not tell us, in the end, which matters are within a sovereign state's discretion.

Additionally, rejection of the assumption plays an important methodological role in the discussion that follows. The shared commitments underlying the current debate over the relevance of and justification for sovereignty are mistaken. In which case, the debate itself must be reconceived. The reconstruction of the debate must begin with a renewed inquiry into the moral nature and limits of state sovereignty. In the end, rejection of the assumption not only casts a doubt on the current debate over state sovereignty it also opens the conceptual space for this inquiry.

References

Beitz, Charles R. 1999. *Political Theory and International Relations*. Princeton: Princeton University Press.

Finnis, John. 2011. *Natural Law and Natural Rights*. Oxford: Oxford University Press.

Hart, H.L.A. 1994. *The Concept of Law*. 2nd ed. Oxford: Oxford University Press.

Hinsley, F.H. 1986. *Sovereignty*. Cambridge: Cambridge University Press.

Hobbes, Thomas. 1996. *Leviathan*. Oxford: Oxford University Press.

Hohfeld, Wesley Newcomb. 1978. *Fundamental Legal Conception as Applied in Judicial Reasoning*. Westport: Greenwood Press Publishers.

Janis, Mark W. 1994. Sovereignty and International Law: Hobbes and Grotius. In *Essays in Honor of Wang Tieya*, ed. R.S.J. Macdonald, 391. Dordrecht: Martinus. Nijhoff.

Kassner, Joshua J. 2009. Completing the Incomplete: A Defense of Positive Obligations to Distant Others. *Journal of Global Ethics* 5 (3): 181–193.

———. 2010. Uti Possidetis: A Philosophical Critique. *APA Newsletters: Philosophy and Law* 10 (1): 2–7.

———. 2012. *Rwanda and the Moral Obligation of Humanitarian Intervention*. Edinburgh: Edinburgh University Press.

Keohane, Robert O. 2003. Political Authority after Intervention: Gradations in Sovereignty. In *Humanitarian Intervention: Ethical, Legal, and Political Dilemmas*, ed. J.L. Holzgrefe and Robert O. Keohane, 282. Cambridge: Cambridge University Press.

Locke, John. 2015. *Second Treatise of Government*. Toronto: Broadview Press.

Morris, Christopher W. 1998. *An Essay on the Modern State*. Cambridge: Cambridge University Press.

Pogge, Thomas. 2008. *World Poverty and Human Rights*. 2nd ed, 29–30. Cambridge: Polity Press.

Raz, Joseph. 1999. *Practical Reason and Norms*, 36. Oxford: Oxford University Press.

Scruton, Roger. 1982. Sovereignty. In *A Dictionary of Political Thought*, 441. London: The Macmillan Press.

Simmons, A. John, and Christopher Heath Wellman. 2005. *Is There a Duty to Obey the Law?* Cambridge: Cambridge University Press.

Steger, Manfred. 2003. *Globalization: A Very Short Introduction*. Oxford: Oxford University Press.

Teson, Fernando. 1998. *A Philosophy of International Law*. London: Routledge.

Wayne Sumner, L. 1986. *The Moral Foundation of Rights*. Oxford: Oxford University Press.

Chapter 3
Assessing the Limitations Inherent to the Notion of Sovereignty Itself

> Sovereignty, as a concept formed the cornerstone of the edifice of international relations that 1648 raised up.[1]

By demonstrating that the accepted view of the legitimate zone of sovereign discretion is dependent on an unjustified assumption, Chap. 2 opened the conceptual space for a freestanding inquiry into the moral nature and limits of sovereignty and sovereign discretion. Our attention now turns to that inquiry. As noted, assessing the moral nature and limits of sovereignty and sovereign discretion unfolds through two distinct but related understandings of the moral limits to the legitimate zone of sovereign discretion. One seeks out the limits to the legitimate zone of moral discretion inherent to the notion of sovereignty itself. The other is grounded in our understanding of the best moral justification for sovereignty. This chapter is focused on the former.

Some may wonder why we should pursue this understanding of the limits to sovereign discretion at all. Why not simply identify and assess the implications drawn from our understanding of the best moral justification for state sovereignty? For one thing, this approach possesses the methodological virtue of being grounded in weak assumptions. It proceeds largely based on the assumption that, as a moral matter, in practice the legitimate zone of sovereign discretion is defined by the rights (and duties) of sovereign states.

In addition, the second approach, pursued in Chap. 4, depends on this one. To clarify, one outcome of the analyses that follow in this chapter will be greater conceptual clarity and understanding regarding the content, scope, and nature of sovereignty. Since one must have an adequate understanding of a norm before they can begin to identify and assess possible justifications, this is necessary for seeking out the best moral justification for sovereignty.

Before proceeding, it may be useful to summarize the conception of sovereignty at issue. As discussed in Chap. 2:

---

[1]Janis (1994), pp. 391, 393.

J. J. Kassner, *Climate Change and Sovereignty*, AMINTAPHIL: The Philosophical Foundations of Law and Justice 10, https://doi.org/10.1007/978-3-030-73578-4_3

1. It is a normative/moral notion;
2. It is a matter of external, rather than internal, sovereignty;
3. It is constituted by a bundle of rights (sovereign equality, territorial integrity, political independence, and collectively, the right of nonintervention);
4. The rights are held by and correlative duties are borne by sovereign states; and,
5. It acts as a second-order exclusionary reason in the practical deliberations of states and the international community.

It is this understanding of sovereignty and the political system founded upon it that serves as the foundation for the inquiry that follows.

3.1 Explanation and Justification of the Methodology Employed

It is important that one understand the methodology being employed and its appropriateness for the task at hand. Having already provided a defense of the non-ideal nature of this inquiry in Chap. 1, the discussion of the methodology here is focused on the approach being taken in this chapter. Simply put, seeking an understanding of the limits to sovereign discretion inherent to the notion of sovereignty itself is an exercise in conceptual analysis, explicating a normative concept and then assessing its implications. As such, the discussion that follows develops an argument grounded in the understanding of sovereignty provided in Chap. 2, and then assesses the implications that argument has for the limits to the legitimate zone of sovereign discretion.

One might question the appropriateness of this approach, pointing specifically to the main assumption underlying this methodology; namely, that it takes the conception of sovereignty developed in Chap. 2 as given. One might object that this either "stacks the deck" in favor of some particular outcome or is unjustified because the conception identified cannot be morally defended.

There are several reasons for pressing on with this line of inquiry. First, there is nothing predetermined in terms of the outcome of the analysis. Assuming that the conception of sovereignty capable of generating the perceived tension has been correctly identified, then the analysis merely determines what implications that conception has for the normative framework of international relations. Second, central to the underlying inquiry is the question, "What are the moral nature and limits of our current understanding of state sovereignty as a foundational moral norm of the normative framework of international relations?" Answering this question must begin with the understanding of sovereignty defended in Chap. 2. One would be justified in objecting if the conception of sovereignty identified was not correct, but not that the methodology is inappropriate. Finally, the argument I offer as the basis for this analysis will be defended in its own right.

3.2 Developing the Argument

As noted, this line of inquiry is grounded in an assessment of the role sovereignty plays as a foundational moral norm of the normative framework of international relations and begins with the delineation of a provisional argument based on the understanding of sovereignty developed in Chap. 2. This provisional argument is then critically assessed, through which the limits to the legitimate zone of sovereign discretion inherent to the notion of sovereignty are identified.

3.2.1 The Provisional Argument

The provisional argument relies on the understanding of sovereignty developed in Chap. 2. There are two aspects to this understanding of sovereignty that, upon reflection, imply limits to the legitimate zone of sovereign discretion. The first is that sovereign states are rights-holders. The second is that the normative framework of international relations is grounded in a commitment to the proposition that the international community is a moral community of rights-holders.

From these two aspects of sovereignty a provisional argument about the limits the legitimate zone of sovereign discretion inherent to the notion of sovereignty can be constructed. The first premise is merely a restatement of what it means for a state to have sovereignty within the normative framework of international relations. Sovereign states are rights-holders and duty-bearers within a community of rights-holders and duty-bearers. The second premise is about the moral relationships that exist between rights-holders within such a community; namely, that the rights possessed by any single right-holder are limited by the rights of others within the community. Together, these premises imply that the rights of sovereign states are limited by the rights of other rights-holding members of the international moral community.

In more formal terms, the provisional argument is as follows:

P1: Sovereign states are rights-holders within the international community, which is a community of rights-holders; and

P2: Within a community of rights-holders, the rights of each individual rights-holder are limited by the rights of other rights-holders.

Conclusion: Therefore, from P1 and P2, the rights of sovereign states are limited by the rights of other rights-holders within the relevant community of rights-holders.

3.2.2 Explicating, Justifying and Revising the Argument

This provisional argument begs several important questions. Regarding P1, as stated, it is unclear who or what constitute the relevant community of rights-holders (and duty-bearers). Thus, raising the question, "Who or what are the other rights-holders within the relevant community of rights-holders?" Much depends on the answer to this question. If there are other rights-holders besides states within the relevant community of rights-holders, then under this argument, legitimate sovereign discretion will be highly circumscribed. If, however, it is states alone, then one might question the reach of the practical implications for such discretion, perhaps concerned that it will be too weak.

In identifying who the rights-holders are within the international moral community, several considerations need to be accounted for. First, the concern here is with moral, not legal, standing. Second, as the methodology is committed to making the least controversial assumptions possible about our understanding of the normative framework of international relations, those identified as rights-holders ought to be widely regarded as such within the disciplines relevant to this inquiry. Finally, another methodological motivation is to presume the broadest range of sovereign discretion.

As to this last point, I take as my interlocutors those who would defend a broad zone of moral discretion for sovereign states. This is motivated by the fact that the objective of this inquiry is to identify the moral limits to sovereignty and sovereign discretion. Consequently, we should assume, as an argumentative foil, a view that implies a broad range of sovereign discretion and thus minimal moral limits to state sovereignty since one way to understand the strength of the arguments presented is the degree to which they can withstand criticism from those who are committed to such a broad range of sovereign discretion. Thus, the methodology being pursued is informed by a commitment to the proposition that the fewer *ex ante* limitations on the legitimate zone of sovereign discretion the better.

As to the proposition that states are the only rights-holders within the international moral community, there are those who continue to hold this view. It, however, conflicts with the consensus found across disciplines and professions that individuals have legal standing as rights-holders within the international arena. To be specific, each individual human being is entitled to have their human rights respected. As such, it would be controversial to offer a conception of the constitution of the international moral community that does not recognize individuals as rights-holders within that community. Perhaps, more importantly, and as will be discussed in greater detail both below and in succeeding chapters, an independent and even less controversial argument can be made that, at the very least, individuals have standing within the international moral community as beneficiaries of moral duties owed to them by states.

One might reasonably ask whether we ought to include multi-national corporations (MNCs) and intergovernmental organizations (IGOs), or nongovernmental organizations (NGOs) as well. MNCs, IGOs, and NGOs are all impactful actors within the international arena; however, there are several reasons for not including any of these entities as rights-holders within the relevant community of rights-holders. First, the ability to influence is different than entitlement. As such, the mere fact that they can act and influence the actions of others in the international arena is not enough to justify a claim to moral standing as a right-holder or beneficiary of duties owed. In addition, the mere fact that many MNC's, NGO's and IGO's have ***legal*** personality does not justify a claim to moral standing.

The fact that the impact of MNCs, NGOs, and IGOs in the international arena does not imply their moral standing as rights-holders or beneficiaries does not deny it either, thus begging the question of their moral standing. In keeping with the methodology being employed, the answer to this inquiry is informed by a commitment to assuming the broadest range of sovereign discretion. In which case, as it would be controversial to assume that they have standing, such collective institutions will not be considered as rights-holders or beneficiaries with moral standing within the international moral community.

One might ask a more foundational question, "Should P1 be accepted at all?" Before answering, it may be useful to identify the nature of P1 as a premise. This helps to identify the sorts of reasons relevant to assessing the acceptability of P1. The proposition that sovereign states are rights-holders within a community of rights-holders, is a descriptive claim. It is a statement of how things are. Though it includes a claim to the normative standing of states in the international arena, it is not a normative proposition. It is not a claim that this is how states ***ought*** to be treated, but that they ***are*** treated this way.

As such, the sorts of reasons relevant to assessing the acceptability of P1 are ones that provide support for the acceptance of such descriptive claims. This does not require that one demonstrate that it is, as a matter of fact, the case that states are rights-holders, but that there are good reasons for accepting this proposition, at least for the sake of argument or analysis. For example, if I claim to have a right, there are several reasons for proceeding as if I have that right, apart from a demonstration that my claim is justified. We may be interested in knowing what it means for anyone to have such a right, or what the best justification is for such a right and what implications this has for the moral nature of my actions. Alternatively, it may be the case that, as a matter of convention, we agree that I have the right, and others may seek to justify their actions towards me based on the existence of the right. In which case, one would want a rigorous analysis of the ostensive right and the normative system in which it is embedded. Neither of these reasons depend on a defense of the proposition that my claim to the right is morally justified.

With this in mind, one reason for accepting P1 is that states often invoke the underlying understanding of sovereignty when they are trying to justify their behavior. As such, P1 is a restatement of the use and understanding of state sovereignty as it is employed by actors within the international moral community. One might object, however, that this use of sovereignty is merely hollow rhetoric intended to obscure more illicit motives. Implicit in this objection is the assumption that sovereignty is about states as rights-holders with standing. What is actually being objected to is the fact that appeals to the protections and permissions that accompany such standing are capable of being illegitimate. Evaluating the legitimacy of such appeals to sovereignty is a separate matter, but it is one that requires a sufficiently clear understanding of sovereignty itself. In which case, you can't assess the legitimacy of an appeal to sovereignty until you have answered the very questions being explored here.

One may, however, be offering a stronger objection, that appeals to sovereignty are ***always*** a pretext. The radical nature of this objection requires a justification; and to be convincing, the justification cannot assume the very proposition it seeks to defend. Consequently, even this objection relies on the initial assumption that sovereign states have standing as rights-holders. To presume otherwise would be to bias the analysis in favor of the conclusion being sought. As such, one methodological reason for accepting P1 is that such radical objections require it.

Another, reason for accepting P1 is that it is inexorably linked to this inquiry. To reiterate, the purpose of this project is to offer a philosophically rigorous understanding of the moral nature and limits of sovereignty, an extant normative principle. In addition, this analysis assesses sovereignty as a foundational norm of the normative framework of international relations. As such, any understanding of sovereignty and its role in the normative framework of international relations other than that expressed in P1 would be focused on something other than the notion of state sovereignty detailed in the previous chapters. In which case, the inquiry would be a vastly different one than the one being pursued.

Finally, is the recognition that sovereignty implies that states are rights-holders within a community of rights-holders a sufficiently complete account of the moral standing of states in the international arena to understand the limitations on sovereign discretion implicit in a system built upon sovereignty? Simply put, no. In Chap. 2, I argued, to adequately understand state sovereignty as a bundle of rights, one must account for the duties correlative to those rights. Failing to account for the moral burdens borne by states in the international arena is not only a failure because the notion is incomplete. That failure does not account for the fact that the correlative burdens borne by states play an important role in limiting the moral discretion of sovereign states. As such, P1 needs to be revised:

> P1*: Sovereign states are rights-holders and duty-bearers within the international community, which is a community of rights-holders and duty-bearers.

One implication of this revision of particular importance is tied to the fact that individuals are also rights-holders within the international moral community. Correlative to those rights are liabilities and duties borne by states; duties owed to individuals. What the exact content of those rights are and the implications they have for the legitimate zone of sovereign discretion will be explored below.

As to P2, unlike P1, P2 is a moral premise. As stated, P2 reads as follows:

> P2: Within a community of rights-holders, the rights of each individual rights-holder are limited by the rights of others.

P2 is the source of the moral turn in the argument. As stated, however, it is poorly suited to that purpose. As such, addressing whether P2 can be justified must wait until P2 has been revised.

The necessary revisions to P2 are based on two considerations. First, P2 must reflect the changes to P1 found in P1*. It would be reasonable to ask why this is necessary; especially considering my assertion that P1 and P2 are categorically different types of premises serving different purposes in the argument under consideration. Notwithstanding, the reasons for altering P1 are replicated in our assessment of P2. The fact that the moral standing of states in the international community is about more than rights alone, that states are also duty-bearers, ought to be reflected in P2. This is merely a recognition of the extant understanding of state sovereignty and the normative framework of international relations.

This leads to the second consideration that needs to be accounted for in our assessment and revision of P2. As stated, P2 is ambiguous. What it means for a rights-holder's rights to be limited by the rights of others within a moral community of rights-holders is left undetermined. To clarify, the relevant sort of limitation is about a limit on the practical deliberations of an agent. The practical deliberations of a right-holding and duty-bearing agent are limited by the rights of and duties owed to other members of the moral community in question.

In light of these considerations, P2 should be revised as follows:

> P2*: In a community of rights-holders and duty-bearers, each right-holding and duty-bearing member's practical deliberations are limited by the rights of and duties owed to others.

Such revisions, however, are not a justification. As such, we will now turn to the task of explaining why one ***ought*** to accept P2*.

The first step in justifying P2* requires an explanation of how the rights of and duties owed to other members of a moral community impact the practical deliberations of members of that community. If we lived hermitic lives, then whatever the content of a right, it would have little impact on the practical deliberations of others. Here, however, we are safe in assuming that rights are a part of the moral framework of a community of rights-holders and duty-bearers, a normative framework that is, by its very nature, social. As such, the content of one member's rights—be they claim rights, liberties, etc.—necessarily make claims on other members in the community. The nature of those claims will determine the impact such rights have on the practical deliberations of other members of the community.

The explanation for how duties borne by members of the community limit the practical deliberations of those same members is likely more intuitive. Nonetheless, there are a variety of moral burdens borne by members of a community of rights-holders and duty-bearers. Here, the focus will be on those burdens correlative to the rights held by members of the community, though this should not be taken to deny there are also primary duties (not correlative to rights) owed to members of the relevant community. The significance of this caveat comes up in a later discussion of the claim that states have duties owed to individuals.

Notwithstanding this, the duties a member of a moral community owes to others within the moral community make claims on the practical deliberations of the duty-bearing member in question. What it means to bear a duty is to have a particular type of moral reason ***must*** be considered in their practical deliberations of the duty-bearing agent. Whether we are discussing the rights of members or the duties borne by members correlative to such rights, how the practical deliberations of members of the community are impacted are the same. The rights held and burdens borne, give rise to moral reasons for or against particular actions or choices. These reasons are to be weighed in the practical deliberations of those agents to whom they apply. That means the practical deliberations of all rights-holding and duty-bearing members within the relevant community will be impacted by the moral reasons generated.

One might be led to reject this premise for Hobbesian reasons, arguing that if the relevant community is a Hobbesian state of nature, then the content of the rights is exhausted by a Hobbesian notion of natural liberty. In which case, there are no limitations. However, under such circumstances, and any others in which the content of the "rights" in question fail to generate moral reasons that serve as limitations on the practical deliberations of others, it would be unreasonable to claim that what is being discussed is a community. Second, as a conceptual matter, even in the Hobbesian situation, we do arrive at an understanding of what limits are generated by the nature of the rights in question, namely none.

One might object that the analysis must be flawed, because in the Hobbesian case it fails to identify any ***effective*** limitations on the practical deliberations of those living in a Hobbesian state of nature. The objection fails to understand the purpose of the premise. It is intended to be a general claim about any community of rights-holders and duty-bearers. The fact that it provides a negative answer in a Hobbesian "community" speaks to its strength. In addition, at this point, the argument is not meant to identify what substantive limitations are implied. That question will be answered when the argument, once justified, is applied to sovereign states as members of the international moral community.

Finally, one might object that since the international arena is a Hobbesian state of nature, there is no point in any further discussion regarding the implications of rights and duties on the practical deliberations of states since such claims are practically meaningless. The objection is based on the proposition that in a Hobbesian state of nature each individual agent has unfettered freedom. Each can act in whatever way deemed best without regard to the effects—harmful or otherwise—on others, with their actions only limited by prudential considerations.

I am unconvinced by this objection, first because there are compelling methodological reasons for rejecting it, many of which have already been provided. In short, the very nature of this inquiry is to assess the moral nature and limits of a norm that is (at least rhetorically) relied upon within the normative framework of international relations. As such, it makes sense to operate under the assumption, even if only for the sake of argument, that the international community is a moral community.

Many may find this response unsatisfying, as it leaves the substantive concerns underlying this objection intact. Thus, as an additional response, the objection relies on it being true that the international community is a Hobbesian state of nature. There is ample reason to be skeptical of this claim. As to the analogy between the international arena and a Hobbesian state of nature, Charles Beitz has convincingly demonstrated the failure of the analogy.[2]

In addition, even if we assume that the international arena shares at least one property with a Hobbesian state of nature, that there is no unitary sovereign to enforce rules, that does not justify the proposition that the normative implications that apply to a Hobbesian state of nature in a parochial setting apply to international relations. The lack of a sovereign is intended to justify the claim that individuals in

---

[2]Beitz (1994).

the state of nature cannot be assured that their rights will be respected by others. It is this lack of assurance that makes following the rules irrational. As such, for the same to be true in the international arena, it must be the case that the lack of a sovereign justifies the claim that states are irrational if they choose to follow the rules of the normative framework of international relations.

As to this claim, Jens David Ohlin has demonstrated that when one takes into account the ability to think beyond the immediate impact of one's actions, and recognize the fact that our interactions with others go beyond single iterations, assurance without a sovereign enforcing the rules is possible.[3] The very aspect of our nature that is taken to support the dire implications of a Hobbesian world, our rationality, can, when enlightened by the recognition of our dependence on others and a longer time horizon, makes adherence to rules—rather than defection from them—rational.

One final point, one might argue that the assurance gained from such an account of Hobbesian enlightened egoism fails to provide an agent with the certainty required to make their rule-following behavior rational. This, however, depends on the proposition that the assurance required for adherence to the rules to be rational must be certain. Even the Hobbesian sovereign would fail that requirement. Rather, the assessment should be based on the probability that the risks are outweighed by the potential benefits. The requisite level of assurance needed is heavily influenced by that evaluation. In which case, it is not whether there is a sovereign or not that matters, it is whether the incentives to cooperate—however they are generated – make such adherence rational.

For some, this may not be enough to convince them of the justifiability of P2*. As such, an example may help to elucidate the justification. Let us imagine a libertarian community in which each member of the community holds a single right, namely, a right to liberty as non-interference. Consequently, when two members of this community come into contact, their interaction is framed, at least from a moral perspective, by this right and the correlative duties they owe one another.

For each member, when they are assessing what they ought to do, they must take the right held by the other and correlative duties they owe into consideration. The right and correlative duties will have an impact on their respective practical deliberations. There will be certain actions that simply aren't permitted and others that are required. Their practical deliberations are not unfettered, they are limited by the rights of others and duties borne to others.

To this point, both premises of the argument have been revised and defended; however, the conclusion has yet to be revised. Beginning with the revised versions of the premises, the argument now runs as follows:

> P1*: Sovereign states are the only rights-holders and duty-bearers within the international community, which is a community of rights-holders and duty-bearers;

[3] Ohlin (2015).

P2*: In a community of rights-holders and duty-bearers, each member's practical deliberations are limited by the rights of other rights-holding members of the community and correlative duties borne by the deliberating agent.

Together then, P1* and P2* lead to the following:

Conclusion*: Therefore, from P1* and P2*, the practical deliberations of sovereign states, as the only rights-holding and duty-bearing members of the international community, a community of rights-holders and duty-bearers are limited by the rights of other sovereign states and correlative obligations borne by the deliberating sovereign states.

The revised conclusion follows from P1* and P2*; consequently, the acceptance of P1* and P2* commits one to accepting Conclusion*.

3.3 Applying the Argument

The Conclusion* provides us with a principled basis for understanding the limitations on the practical deliberations of states inherent to the concept of sovereignty. This understanding serves as the basis for understanding the limits to the legitimate zone of sovereign discretion inherent to the notion of sovereignty itself.

One might question this connection, arguing that it is not obvious how the impact state sovereignty has on the practical deliberations of states limits the legitimate zone of sovereign discretion. To be more specific, the objection would be based on the assertion that Conclusion* can be used to provide an understanding of the impact the rights held by and duties borne by sovereign states has on the practical deliberations of sovereign states, but does not directly address the matter of sovereign discretion.

In response to this concern, the following is a brief explanation of the connection between the limitations on the practical deliberations of states and limits to the legitimate zone of sovereign discretion. As noted, practical deliberation is about the reasoning an agent employs when deciding what they ought (all-things-considered) to do. If the agent's practical deliberations are unfettered, then they have wide discretion over what reasons they consider, the weight they give to them, and the action they choose. If, however, an agent's practical deliberations are encumbered by any variety of limitations, their discretion is limited. Ultimately, limitations on an agent's practical deliberations are limitations on the agent's discretion.

3.4 Conclusion

The final, and perhaps most substantive step, in fleshing out the implications this argument has for how we ought to understand the legitimate zone of sovereign discretion, requires that we assess the implications Conclusion* has for the practical deliberations of states in the international arena. It is this understanding that forms the basis for the limits to the legitimate zone of sovereign discretion inherent to the notion of sovereignty. The conceptual work has already been done and we are left with the task of stating explicitly what the practical implications of Conclusion* are.

Simply put, the legitimate zone of sovereign discretion is limited by the constraints on the practical deliberations of states inherent to the notion of sovereignty. Which, according to Conclusion*, means that the practical deliberations of states are constrained by the rights held by and duties owed to other members of the international moral community. As such, the legitimate zone of sovereign discretion is also limited by the rights held by and duties owed to other members of the international moral community. The implications of this understanding of the limits to the legitimate zone of sovereign discretion are further explored in Chap. 8, wherein it is determined whether the challenge of climate change falls within the limits to the legitimate zone of sovereign discretion inherent to the notion of sovereignty.

References

Beitz, Charles. 1994. *Political Theory and International Relations*. Princeton, NJ: Princeton University Press.

Janis, Mark W. 1994. Sovereignty and International Law: Hobbes and Grotius. In *Essays in Honor of Wang Tieya*, ed. R.S.J. Macdonald. Dordrecht: Martinus Nijhoff.

Ohlin, Jens David. 2015. *The Assault on International Law*. Oxford: Oxford University Press.

Chapter 4
Assessing the Limitations from the Best Available Justification for Sovereignty

> The conflicting allegiances of Europe had contributed to the terrible toll of confusion, death, and destruction from 1618 to 1648. In the mid-seventeenth century many Europeans sought a simpler, and it was hoped, safer set of loyalties.[1]

> [I]n a world of states, sovereignty protects one against outsiders trying to topple the government or to set up a puppet regime or to impose their views of what is good and right – hence the particularly strong attachment of countries recently liberated from colonial rule to the principle of nonintervention.[2]

The discussion now turns to identifying and assessing the limits to the legitimate zone of sovereign discretion implied by the best available moral justification for state sovereignty. The underlying approach involves a methodology comprised of two distinct, but interrelated, steps. To begin, it is necessary to identify the best available moral justification for sovereignty. Only then are we able to embark upon the second step and assess the limits to the legitimate zone of sovereign discretion implied by that justification.

4.1 Preliminary Matters

Before turning to that discussion, to avoid misunderstandings, it is necessary to address several methodological issues. For one, though both discussions in **Chap. 3** and this chapter identify moral limits to sovereign discretion, there is a significant methodological difference between the two approaches. In **Chap. 3**, there was no need to inquire about the moral value of or moral basis for sovereignty, rather the inquiry focused on the conceptual implications that follow from the recognition of

---

[1]Janis (1999), p. 159.

[2]Hoffman (1996), p. 19.

J. J. Kassner, *Climate Change and Sovereignty*, AMINTAPHIL: The Philosophical Foundations of Law and Justice 10, https://doi.org/10.1007/978-3-030-73578-4_4

the fact that the conventional notion of sovereignty is of sovereignty as a right. Here engaging in such moral inquiry is unavoidable.

Another methodological matter to be addressed is related to the fact that the moral justification being sought takes the conventional understanding of sovereignty as given. One may contend that this amounts to little more than an apology for the various and sundry ways in which the conventional understanding is employed within the international arena. The objective here, however, is not to identify and defend a justification that renders the conventional conception immune from criticism. Rather, the purpose is to identify and defend the ***best available*** moral justification for the conventional understanding of sovereignty, and then assess the implications this has for the limits to the legitimate zone of sovereign discretion.

The virtue of this approach is that it takes the belief in the moral nature of sovereignty seriously without accepting all of the implications that the conventional understanding of sovereignty as a moral norm is taken to have. The extent to which the limits to the legitimate zone of sovereign discretion implied by the best available moral justification for sovereignty are inconsistent with accepted practices or beliefs about the moral nature and limits of state sovereignty, so much the worse for those conventional beliefs and practices.

It may be useful to provide a brief explanation of the methodology being employed. The analysis begins by identifying a variety of different candidate justifications. Those candidate justifications are then evaluated based upon a set of conditions that any viable justification must meet. The identification of those conditions is influenced by L. Wayne Sumner's work on moral rights.[3] For Sumner, understanding the moral foundations of a norm should begin with the identification of the constitutive elements of the norm in question, namely, their content and scope.[4] The constitutive elements then serve as the existence conditions any viable candidate justification must be able to accommodate.[5] If an account is incapable of doing so, then it is not viable as a justification for the right in question.[6]

In addition to the content and scope, I have contended elsewhere that the existence conditions of a norm should also include the identification of the ***nature*** of the norm in question.[7] By "nature" I mean the impact the norm has on the practical deliberations of those agents to whom the norm applies. As to why this should be so, norms are practical in the sense that they are intended to guide us in our choice of actions; consequently, to understand a norm it is essential to understand the role the norm plays in an agent's practical deliberations.

Despite this, one might wonder why this methodology grounded in Sumner's understanding of the moral foundations of rights is being pursued. Though a complete defense is not possible here, there are at least two compelling reasons in

[3]Sumner.

[4]*Ibid.*

[5]*Ibid.*

[6]*Ibid.*

[7]Kassner (2012).

its favor. First, has been mentioned, the larger project is explicitly focused on the conventional understanding of sovereignty. As such, any effort to identify the best justification for sovereignty, should be tied to the constitutive elements of the notion of sovereignty at issue. Second, as will become apparent, there are numerous candidate justifications for this understanding of sovereignty. The approach being adopted provides a principled basis for distinguishing between those that ought to be taken seriously and those that are not worth further consideration.

There is one final preliminary matter that needs to be discussed. Namely, what are the constitutive elements that must be accommodated? Simply put, the constitutive elements that will serve as the existence conditions are those found in the explication of sovereignty in **Chap. 2**:

Scope: States are holders of the rights attendant to the possession of sovereignty, and states and the international community, understood as a collection of duty-bearing states, are the bearers of the correlative moral burdens.

Content: The content includes the rights to territorial integrity, political independence, and sovereign equality, which, collectively, constitute the right of nonintervention. In Hohfeldian terms this implies an immunity against interference with territorial integrity and political independence, an entitlement to be treated as an equal within the international arena ***and*** a power to alter the normative relationships of outsiders as they relate to the territorial integrity and political independence of the rights-holding state.

Nature/strength: Sovereignty (through the right of nonintervention) gives rise to second-order exclusionary reasons against those reasons for action the fulfillment of which would require violation of a sovereign state's right of nonintervention. The inability of a justification to accommodate this element would amount to a failure to accommodate the role sovereignty plays in the normative framework of international relations.

4.2 Identification and Evaluation of Arguments

As noted, the methodology being employed depends on our ability to identify the best available moral justification for sovereignty. Doing so requires that one first identify the alternative candidate justifications available. Once those candidate justifications have been identified, we can then assess the extent to which any such justification(s) is(are) able to accommodate the existence conditions identified.

4.2.1 Responding to the Objection That the Inquiry Is Moot

Before proceeding, there is one objection to the need to identify candidate justifications that needs to be addressed. Namely, one could argue that this inquiry is moot because it is a ***social fact*** that political communities with a defined contiguous territory, the ability to provide internal and external security, and are recognized by the international community as sovereign are sovereign. The normative framework of international relations attributes certain rights to political communities that possess sovereignty. Thus, this is all there is to understanding sovereignty and its justification, and there is no further need for a moral justification.

This objection depends on it being true that ***either*** conventional rights are equivalent to moral rights, in which case the mere existence of a conventional right is sufficient justification; ***or,*** that justifications based on the rules of a conventional system are equivalent to justifications based on moral argument, in which case the fact that the rights in question are justified by the conventional system suffices. Neither proposition is true. There are, at least, three differences between conventional and moral rights relevant to this objection.

First, conventional rights that are amoral, and in some cases immoral, are possible. There is no necessary connection between conventional rights and morality. On the other hand, moral rights are, by definition, moral. To clarify, take an example based on H. L. A. Hart's defense of the separability thesis against the charge that the existence of legal rights implies a necessary connection between law and morality. One need only consider the rights found in games to show that rights devoid of moral content can exist.[8] In (European) football, if a field-player touches the ball in the 18-yard box with their hand in their defensive end, the opposing team has a ***right*** to a penalty kick. If it is not awarded, we are justified in criticizing the referee. There is nothing necessarily moral about this right. On its own, the right to the penalty kick is an amoral conventional right.

Second, conventional rights and moral rights occupy distinct normative spheres of influence. The practical relevance of moral rights as justifications and reasons for action, when compared with conventional rights, is broader. Conventional rights are practically relevant within the conventional systems in which they operate and not outside of that context. Returning to our example, the right to a penalty kick only makes sense during a football match. Moral rights, on the other hand, are relevant in a more general sense, often serving as the basis for criticism of, or justification for, the conventional rights that constitute the internal normativity of a conventional system.

Third, and finally, there is a distinction in how one assesses the validity of conventional versus moral rights. The validity of a conventional right is tied to its status as a social fact. If, within a conventional system, the agents whose conduct is governed by the conventional system endorse a right, then it is a valid right of the system. As such, the validity of a conventional right ***as a conventional right*** is a

[8] Hart.

matter of approval within the system. This is not a moral justification. The validity of a moral right, on the other hand, depends on a moral justification, which is external to any conventional system. In the end, moral rights and conventional rights are not equivalent; consequently, the first proposition fails.

One might concede this point yet argue that the right is justified by the role it plays in the conventional system. This form of justification depends on it being the case that a conventional system can provide the requisite moral justification. This is only true, however, if justifications based on the connection a right has to a conventional system are equivalent to moral justifications.

Thus, we are led to consider this second proposition, and assess whether there is such an equivalence. It ought to be rejected, first because, like the discussion of conventional rights, justifications based on the role a right plays in a conventional system are conditional in a way that moral justifications are not. The relevance of a conventional justification depends on, the matter or claim being justified being internal to the conventional system itself. Moral justifications are more generally applicable and are justifications based on moral reasoning external to any such conventional system. Though conventional systems can be the object of moral assessment, the inverse is not true.

Another significant difference between justifications based on moral argument and those based on a conventional system harkens back, again, to H.L.A. Hart's defense of the separability thesis. Within a conventional system there is no necessary connection between the rules of the system and morality. This does not mean that there can't be a conventional system that, as a matter of contingent fact, is based on morality, but that does not make conventional justifications equivalent to moral justifications. It may be true for any particular conventional system that it is morally justified, in which case it may be able to provide a moral justification for the conventional rights of the system, but that is because of the underlying moral justification of the conventional system, not because it ***is*** a conventional system.

To elucidate, returning to our example, the awarding of penalty kicks plays an integral part in the game of football. There are numerous reasons for their inclusion. Some of which ***may*** be moral; however, the mere fact that there may be a conventional justification for penalty kicks tied to the role they play in the game does not mean that they are morally justified. We need to inquire further into the reasons why penalty kicks are included to determine whether there is anything more than a conventional justification. It may be that the inclusion was intended to increase scoring by incentivizing attacking players to be more aggressive while deterring defending players from responding aggressively near the goal. A conventional justification leaves the question of a ***moral*** justification open.

Ultimately, whether conventional rights and justifications can be morally justified begs the very question being pursued here—what possible moral justifications exist for the conventional rules at issue. In other words, social facts are not moral facts and conventional justifications are not moral justifications. If one is seeking a moral justification for a conventional rule, one cannot avoid the necessity of looking outside the conventional system and its norms.

4.2.2 *Identification and Evaluation of the Candidate Justifications*

The discussion can now return to identifying the candidate justifications. Recall, this inquiry is motivated by the dearth of scholarly discussion on the moral nature and limits of sovereignty. Consequently, one might wonder how candidate justifications will be identified when such justifications are, presumably, unavailable. Rather than constructing arguments from whole cloth, I am adopting arguments found within the scholarly literature offered as moral justifications for the right of nonintervention. Though intended as justifications for a state's right to be free from intervention and not sovereignty *per se,* such arguments are offered as justification for the most morally significant aspect of state sovereignty.

4.2.2.1 Arguments based on the Domestic Analogy

Several arguments employ some form of the domestic analogy. Charles Beitz, though a critic of their use, offers a succinct statement of the role domestic analogies play in international political theory. As Beitz notes, "like persons in domestic society, states in international society are to be treated as autonomous sources of ends, morally immune from external interference, and morally free to arrange their internal affairs as their government sees fit."[9] Arguments from liberty and arguments from anti-paternalism are two of the most common forms of argument within international political theory that rely on the domestic analogy.

Notwithstanding this, and before those arguments are addressed, there is one form of analogical argument that I want to address briefly, both to identify it and to explain why it won't be considered further. It is a Hobbesian argument that needs to be addressed because it could be confused with a different argument from Hans Morgenthau in his discussion of the practical relevance of morality to the normative framework of international relations, and could also be confused with a compelling argument recently made by Jens David Ohlin regarding the rationality of rule-following in the international arena.[10]

The argument begins by recognizing that in a Hobbesian state of nature there is no sovereign to enforce rules, including moral rules. As such, according to Hobbes, in this pre-social condition individuals possess absolute freedom. This does not, however, lead to an anarchist utopia; rather, it creates a world of perpetual conflict—a war of all against all.[11] Under such circumstances, it would be irrational for any individual to abide by rules that constrained their behavior, because doing so would

[9]Beitz (1999), pp. 65–66.

[10]Ohlin.

[11]Hobbes, pp. 82–84.

put the individual's self-preservation under threat. Therefore, in the state of nature, the only norm that ought, rationally, to guide an individual is prudence.

There is no sovereign in the international arena to ensure that states follow the rules, be they legal, moral or otherwise. This has led many scholars to claim that states in the international arena exist in a Hobbesian state of nature. By analogy, this means that states are ***absolutely*** free, but also that each is an ever-present threat to all others. If the analogy holds, then the only norm that makes rational sense for states to follow within the international arena is prudence.[12]

Charles Beitz has refuted this claim,[13] but, for argument's sake, I will set his rejection aside. Notwithstanding this, those seeking to justify sovereignty in light of this argument argue that, because international relations is a system of multiple iterations over time rather than single interactions, states ought, as a matter of prudence, comply with the normative demands of state sovereignty. As such, it can be ***prudent*** for states to respect sovereignty and for the normative framework of international relations to be built around that commitment. Versions of this argument are compelling demonstrations, *contra* many realists, of the fact that rule-following can be rational, even in a Hobbesian state of nature.[14] What such arguments do not do, however, is provide a ***moral*** justification for such rules.

In addition, as Hobbes concludes with regards to the actions of states, "every commonwealth, (not every man) has an absolute liberty, to do what it shall judge (that is to say, what that man, or assembly that representeth it, shall judge, most conducing to their benefit."[15] In other words, the Hobbesian argument justifies the prescription that a state ought to pursue the course of action that is in its best interest. In which case, if the circumstances demand it, a state may, out of prudence, be rationally compelled to disregard any rule that interferes with its interests. Consequently, the justifications produced are practical and only contingently support respect for sovereignty. I will not consider this argument further.

Another, more promising, form of analogical argument is grounded in John Stuart Mill's understanding of the value of individual liberty. Mill argues, "[i]n the part [of an individual's conduct] which merely concerns himself, his independence is, of right absolute. Over himself, over his own body and mind, the individual is sovereign."[16] Thus, respecting individual liberty requires a broad right of noninterference, especially when the individual's actions are self-regarding. Assuming Mill is correct and that states in the international arena are like individuals in domestic society, many have argued that this implies a right of nonintervention for states.[17] If true, sovereignty implies a similarly broad right against intervention.

---

[12]See, Benn and Peters (1965).

[13]Beitz (1999), pp. 36–40.

[14]Ohlin (2015).

[15]Hobbes, p. 142.

[16]Stuart Mill (1991), p. 14.

[17]Benn and Peters, pp. 429–431, and Vincent, p. 345.

Another form of argument relying on the domestic analogy is based on the value of anti-paternalism. Since there are competing conceptions of the good and we can't be sure which is correct, William E. Hall, argues that a principle of anti-paternalism implies that individuals should be free to choose what constitutes the good for themselves.[18] The argument proceeds by assuming that states in the international arena are like individuals in domestic society and that there are competing conceptions of the (political) good from which states can choose.[19] Therefore, by analogy, the principle of anti-paternalism should govern the interactions of states in the international arena.[20] Hall concludes that respect for a state's sovereignty is dictated by a commitment to anti-paternalism, wherein other states and the international community must refrain from interfering in the internal affairs of a state as such interference would require substituting an outsider's conception of the good for the state's.

It remains to be seen whether such analogical arguments can accommodate the constitutive elements of state sovereignty. Notwithstanding this, for an analogical argument to work, the following must be true. First, the grounding proposition must be true. Second, it must be the case that the elements of the claim being defended are sufficiently similar in morally relevant ways to those in the grounding proposition. Finally, even if the analogy holds, it must generate the conclusions constitutive of the moral claim being defended, in this case the content, scope, and nature of sovereignty.

Here, let us assume the grounding proposition is true, that respect for individual liberty requires that others not interfere with an individual's self-regarding choices. Next, we must ask whether, within the proposition being defended, the elements morally relevant to the defense of the moral claim at issue—that state sovereignty can be justified by an analogy with individual liberty or anti-paternalism—are sufficiently similar to the grounding proposition. This requires states to be sufficiently similar to individuals in terms of their respective moral nature and worth.

In answering, we must first explain what it is about individuals that implies a right of noninterference. I am not able to engage in a full explication and assessment of this grounding proposition here; however, as the analysis being pursued is primarily concerned with a different inquiry, and assumes that this proposition, that individuals have a right of noninterference, is justified, the justification offered for that proposition is in the form of an explanation of the argument that, in my considered judgment, is most convincing.

As such, the most compelling argument for the grounding proposition begins with a claim that, due to their liberty, individuals are (or, at least ought to be treated as if they are) valuable in-and-of-themselves. It is then argued that their self-regarding choices are, by implication, intrinsically valuable. Interference with such choices fails to respect this intrinsic value, and as a consequence, are *prima facie*

---

[18]Hall (1924), p. 347.

[19]For discussion, see Beitz, p. 88.

[20]Hall, p. 347.

morally wrong. Consequently, respect for the intrinsic moral worth of individuals and avoidance of the moral wrong that comes from interference implies a right of noninterference.

So, are states in the international arena sufficiently similar to individuals in domestic society? In a conventional sense, states are ***treated*** like individuals in domestic society, and state sovereignty ***is thought of*** in ways that are similar to the way individual liberty is thought of in domestic society. The fact of such treatment is not, however, a justification. Rather, it begs the very question being asked, whether that conventional treatment can be justified.

As such, we must return to the question of whether, as a moral matter, states in the international arena are like individuals in domestic society. As the justification for the grounding proposition is that individuals are intrinsically morally valuable, it must be the case that states can be understood to be intrinsically morally valuable as well or, at least, that we ought to treat them so.

Assessing this proposition requires that we understand what it means for a state to be valuable in-and-of-itself. For one thing, this would mean that states have moral standing ***and*** moral worth separate from their relationship to any other moral agents, independent of any benefits they confer on others, including those that inhabit the state itself. As to the former, I doubt that such independent moral standing can be justified. With that said, to protect against confusion, I will entertain two arguments that are intended to justify the independent moral standing of states. One might argue that states are the apex of human political development, the object towards which our political history has been aiming. Such arguments are historicist in nature and are belied by the facts. In addition, these arguments don't justify the independent moral standing of the state; they assume it.

Another argument offered as a justification for the independent moral standing of the state originates in the work of Hobbes. For Hobbes, the social contract that creates the state is grounded in an alienation contract. In other words, the state is created when individuals literally ***give*** their rights to the state. The state, once created, holds these rights as its own.

One might object to this interpretation of the Hobbesian social contract, noting that it is more complicated, with individuals retaining certain rights against the state. In which case, there are limitations on state behavior that are tied to the state's treatment of individuals, extremely narrow limitations, but limitations nonetheless. As an exegesis of Hobbes, this is likely correct, but this is not an exegesis of Hobbes. More importantly, the objection as stated would render the argument from the Hobbesian social contract incapable of justifying the independent moral standing of the state, as the standing of the state would remain contingent on its treatment of individuals.

There are several reasons for rejecting this Hobbesian argument. First, it is simply untrue. There has never been any agreement that would justify the claim that the inhabitants of any modern state have voluntarily alienated their rights. Second, one might point to the fact that the Hobbesian social contract is not to be taken literally. It is a hypothetical contract based on the claim that such agreement is rationally mandated by the threat of the state of nature. This may arguably explain the rational need for a state, but it does not justify the proposition that states have independent moral standing.

One might object that, even though the independent moral standing of the state can't be justified, that we should ***act*** as if it has such standing for a variety of instrumental reasons. If one takes this tack, then one has abandoned the claim that states are intrinsically morally valuable. In response, it could be argued that, not only should we treat states as if they have independent moral standing, we should ***act*** as if they have intrinsic moral worth as well. This does nothing to resolve the problem, but merely replicates the question at another level. In either case, it is an admission that the moral worth of the state is instrumental not intrinsic, not valuable in the same way individuals are.

Setting aside these criticisms, let us assume, for the sake of argument, that the analogy between states in the international arena and individuals in domestic society holds true. This would justify the proposition that states are rights-holders in the international arena, thus accommodating the scope of sovereignty. It remains to be determined whether the content of state sovereignty can likewise be accommodated.

Returning to the grounding proposition, but with a more complete understanding of its implications, the only choices and actions against which interference is prohibited are those that are self-regarding. Actions that impact other moral agents are not protected by the ***right*** of noninterference. ***All*** choices by a state, impact other moral agents, perhaps most importantly, the inhabitants of the state itself. This has implications for the ability of the analogical arguments to accommodate both the content ***and*** nature of sovereignty. Since none of the actions of the state are self-regarding, they would not be entitled to the protections of the right of nonintervention under such an argument. In addition, since the right does not apply, whatever respect morality demands we give to sovereignty in our practical deliberations under such circumstances, we need not treat it as a second-order exclusionary reason, thus failing to accommodate the nature of sovereignty.

Based, however, on the assumption that states have independent moral standing and intrinsic value, one might argue that the limitation just discussed is a moot point, that a state's analogical rights, because of the state's independent moral standing and intrinsic value, are inviolable. This objection conflates ***intrinsic*** with ***absolute*** value. For the objection to succeed, states must have absolute, and not merely intrinsic, value. The fact that ***A*** has a separate moral identity and is intrinsically valuable does not imply that interference with ***A's*** actions and/or choices is impermissible. Rather, it merely defines the way we must understand ***A's*** value within our practical deliberations. To respect ***A*** requires more than simple consequentialist reasoning. In addition, we are concerned with choices by ***A*** that are not self-regarding; consequently, they fall outside the ambit of the right, even if we assume the intrinsic value of the state.

Domestic analogies may be useful as explanatory devices; and, in international political theory, they are no doubt helpful in explaining the role state sovereignty plays in the normative framework of international relations. Such analogical arguments, however, cannot accommodate the constitutive elements of sovereignty. Therefore, analogical arguments fail as moral justifications for sovereignty.

4.2.2.2 Non-Analogical Arguments

There are several non-analogical arguments worth considering. The first is an argument made by Michael Walzer. Walzer relies on an argument from political self-determination found in Mill's "A few words on nonintervention."[21] Though there is some disconnect between Walzer's understanding of Mill's argument and the argument Mill himself makes, both argue that political self-determination requires that a people seeking political freedom achieve it on their own; consequently, outside interference in such efforts is unjustified as it would interfere with the people's ability to effectively attain a lasting political freedom.

As Walzer argues, "[t]he members of a political community must seek their own freedom, just as the individual must cultivate his own virtue. They cannot be set free, as he cannot be made virtuous, by any external force. Indeed political freedom depends upon the existence of individual virtue, and this the armies of another state are most unlikely to produce. . . ."[22] A such, the "state is [. . .] the arena within which self-determination is worked out from which, therefore, foreign armies have to be excluded."[23]

It may help if we present this argument in more formal terms:

Premise 1: Political freedom (political self-determination) is valuable;

Premise 2: Lasting political freedom requires that a people have certain virtue and attitudes;

Premise 3: The virtues and attitudes necessary for lasting political freedom cannot be attained with the help of outsiders.

Conclusion: Therefore, outsiders must be prevented from interfering in a people's struggle to attain political freedom.

Thus, according to Walzer, to respect a people's political self-determination, the people in question must be allowed "to seek their own freedom" without interference from others.

Looking first to whether Walzer's argument can accommodate the scope of state sovereignty, under this argument, the right of nonintervention is tied to political freedom, and the value of political freedom is attached to ***a people***. As such, whether this argument can accommodate the scope of sovereignty depends on the relationship between states and peoples.

Simply put, a people is not equivalent to a state; they constitute morally distinct categories.[24] As such, an explanation of how the two are related is needed. For Walzer, the connection is based in a presumption; namely, that there is a "fit" between a people and the state within which that people resides.[25] The justification

---

[21] Stuart Mill (1867).

[22] Walzer (2000), p. 87.

[23] Walzer (1980), p. 210.

[24] Rawls (1999).

[25] Walzer (1980).

for this presumption depends on it being true that we (outsiders) can't know with sufficient certainty that there is ***no*** fit.

This presumption is based on several empirical assumptions. First, it must be the case that outsiders don't have the necessary epistemic access to the relationship between a people and a state, to be able to assess the nature of that relationship. This is belied by the fact that in a globalized world in which the telecommunications revolution and rise of social media has fundamentally altered our epistemic relationship to others, we ***can*** know, in real time, what is happening thousands of miles away in distant corners of the globe.

Second, Walzer's argument also depends on each state being occupied by a single people. If we take ***a people*** to be a moral category, then nearly every modern state is made up of multiple peoples.[26] In which case, one cannot claim that state sovereignty is protective of political freedom for a single people without offering some argument as to which people and why. If Walzer means something different, then he owes us an argument, which can't simply be an assumption based on a tautology—that a people is the collective political community comprised of the inhabitants of a state. The argument must explain what it is about a people that makes them a moral entity and not merely a collection of individuals that happen to reside near one another in a contiguous territory.

Assuming there is an argument that can explain the relationship between a people and the state, that doesn't explain why states should be rights-holders. For Walzer, the explanation is grounded in his claim that the state is where the virtues and attitudes necessary for lating political freedom are cultivated. Specifically, through their struggle for political freedom, a people cultivates such virtues and attitudes.

In assessing this claim, there are, at least two ways to understand the relationship between sovereignty and a people's political freedom, one weak and one strong. Under the weaker understanding, respect for state sovereignty, though not necessary, is an effective means to cultivating the virtues and attitudes necessary for a lasting political freedom. This interpretation of the argument is similar enough to the instrumental argument from political self-determination, that I will not consider it separately here, rather it should be considered alongside the discussion of the instrumental justification for sovereignty discussed below.

Under the strong understanding of the relationship, state sovereignty is ***necessary*** for the cultivation of the virtues and attitudes required for lasting political freedom. As such, the standing of the state as a rights-holder is dependent on it being true that sovereignty is necessary for the cultivation of the relevant virtues and attitudes. We should be skeptical of this proposition. For one thing, it is unclear why lasting political freedom depends on ***particular*** virtues and attitudes. It would seem just as likely that a wide variety of conflicting attitudes and virtues could provide for a lasting political freedom. Even if it is the case that particular virtues and attitudes are necessary, it is unclear why sovereignty is necessary to cultivate them. It is certainly

[26]For a broader understanding of what constitutes a "people" see Rawls (1999). See also, Margalit and Raz (1990), pp. 439–461. And see Heath Wellman (1995), pp. 142–171.

logically possible that a people could develop such virtues without precluding those who may wish to assist them in their efforts from providing that assistance.

There are also historical examples that undermine the existence of a necessary relationship between lasting political freedom and the need for a people to struggle without assistance to achieve it. If such a strong relationship exists, then we should be skeptical of the ability of Canadians, Australians and others who achieved their political freedom through patience, diplomacy and ***the grace*** of the British government to maintain their political freedom over time.

In any event, Walzer doesn't offer an argument for the proposition. What is offered is empirical speculation that the cultivation of such virtues requires an institutional structure governed by sovereignty and the right of nonintervention, or that such virtues would be undermined in an normative framework of international relations that permitted outsiders to assist those seeking political freedom. If the relationship is based on an empirical claim, then evidence needs to be mustered to that effect. In addition, empirical claims are contingent claims, and do not support a norm of the sort being discussed here.

Walzer's argument also implicitly assumes that it is ***possible*** for a people to attain political freedom through unaided struggle. Despite the fact that even the American colonists required the assistance of France to obtain its freedom from the United Kingdom, for the sake of argument, I will concede that this may have been true in the nineteenth Century when Mill published on the matter, but it is not clear that this is true today. One need only think of the plight of present-day Syrians to understand the capacity for violence that a despotic regime can visit upon its own people to maintain control. What Walzer makes available to many of those seeking political freedom today is annihilation or subjugation, not political freedom. It is reasonable to believe that, in most cases, struggle is important to the ***appreciation*** of political freedom. But, without further argument, it is not clear why unaided struggle is ***necessary***, and for that reason, why states must be rights-holders to make lasting political freedom possible.

Notwithstanding this, even if we assume that Walzer's argument can accommodate the scope of sovereignty, it remains to be determined whether it can accommodate the content? Specifically, can it accommodate the right of nonintervention? This would require a demonstration that the right of nonintervention is necessary to the cultivation of the virtues and attitudes that are likewise necessary for lasting political freedom. It has already been noted that it is logically possible to have lasting political freedom with a different mix of virtues and attitudes. In addition, in the case of the ability of a people to attain political freedom through unaided struggle, the assumption likely conflicts with reality. Unaided resistance and struggle in the pursuit of political freedom against well-armed and powerful despots is a fool's errand.

Consequently, sovereignty and the right of nonintervention could reasonably be expected to interfere with a people's ability, if not right, to attain lasting political freedom. In the end, this is an empirical claim, not a conceptual one. For that reason alone, we are justified in concluding that there is no necessary relationship between sovereignty and either the cultivation of the virtues and attitudes necessary for lasting political freedom or lasting political freedom itself.

Returning to the Millian argument underlying Walzer's account, the normative breadth of the justification is limited because the objective is the protection of a people's right to political self-determination, which is, in turn, linked to individual freedom. Individual freedom may, however, conflict with a people's political self-determination. This potential conflict undermines the ability of Walzer's argument to accommodate either the content or the nature of sovereignty.

As to the content, since a people's political freedom is based on individual self-determination, if the exercise or protection of that people's freedom undermines individual self-determination, it is not clear why a people's political freedom should take priority, which it must if the content of sovereignty is to be accommodated. Regarding the ability of the argument to accommodate the nature of sovereignty, due to the potential moral conflict between a people's political freedom and individual self-determination, we are presented with an open question as to which moral commitment should take precedence. Thus, we are faced with competing reasons to be weighed against one another in the practical deliberations of states, not a decisive argument for treating sovereignty as a second-order exclusionary reason. As such, Walzer's argument is incapable of accommodating the nature of sovereignty.

4.2.2.3 Non-Analogical Arguments: S. I. Benn and R. S. Peters's Argument

The second non-analogical argument to be considered comes from S. I. Benn and R. S. Peters.[27] Their argument is based on the proposition that the members of a political community choose a particular conception of the political good. As Benn and Peters state, "[n]ow nation-states are themselves associations of individuals with their own common interests and aspirations, expressed within a common tradition."[28] In other words, a political community's choice of a political good is representative of the common interest of that community and an expression of their collective will.

Sovereignty is argued to be necessary to show respect for this choice of political good, and failures to respect the sovereignty of a state are argued to be a moral affront to the individual members of the political community. Benn and Peters put forward two distinct moral grounds to support the claim that the individual members of the community have been wronged, one Kantian and the other Millian.

As to the Kantian argument, Benn and Peters contend that "[w]hen one state meddles in another's affairs, the nationals of the victim are rarely considered on the same footing s those of the interfering state. The latter treat them rather as means for its own ends."[29] In this argument, the value of sovereignty is tied to its ability to

[27]Benn and Peters, p. 361.

[28]*Ibid.*

[29]*Ibid.*

ensure that the members of the political community are respected as ends-in-themselves and not used merely as means.

The success of this argument depends on at least one improbable proposition; namely, that sovereignty is necessary to ensure that individual members of a political community are respected as ends-in-themselves. Sovereignty is neither necessary nor sufficient for this purpose. Benn and Peters' own argument recognizes this possibility. When it matters most (i.e. in cases of human rights violations, massive political corruption, political violence, etc.) outsiders are often more committed to the humanity of the individual members of a political community than are the leaders of the community in question. States often invoke their sovereignty to shield their abuses from outside interference, not to protect their citizens as ends-in-themselves. I will not consider this argument further.

The alternative argument is based on Mill's defense of anti-paternalism and noninterference in the self-regarding choices of individuals. Unlike the arguments previously discussed, this argument does not depend on the domestic analogy. Rather, it is a consequentialist argument. For Mill, individuals should be free to make choices for themselves because they know and care for their interests best.[30] Such free choices are more likely to satisfy their interests than are choices imposed upon them, even if those choices are imposed on them by others who have the best of intentions. If true, then non-interference in such choices is likely to lead to the greatest overall utility.

For Benn and Peters, "it is doubtful how far any state would be justified in deciding what was best for other peoples, just because it is so difficult to know the needs and attitudes of the people one aspires to help."[31] Beitz, in discussing this argument, notes that it is grounded in the proposition that "intervention in a state's internal affairs cannot be justified on paternalistic grounds because the intervening state is unlikely to be impartial and because, in any event, a state is more likely to know its own best interests than any other state."[32] This argument depends on it being true that if one refrains from interfering in the choices of a political community, the common interests of that political community are more likely to be served.

This argument thus depends on it being true that an agent's interests are best served when that agent is free to make choices for themselves. If true, an agent should have a right to noninterference. To the extent that a political community is an agent with interests, then, under this argument, such interests are best satisfied when each political community is free to govern itself and pursue policies of its own choosing.

Moral consequentialism, however, is not merely means-ends reasoning. It is means-ends reasoning directed at a moral end. For example, utilitarianism, is dependent on it being true that utility (or happiness) is ***the*** morally valuable end towards which all action should be directed and by which all action is morally

[30]Mill (1991).

[31]Benn and Peters, p. 362.

[32]Beitz (1999), p. 84.

evaluated. Pursuant to this, one caveat the must be added, is that the interests of a political community must be morally valuable to serve as the basis for a consequentialist justification. Though it is certainly possible that the interests of a political community are morally valuable, this is not necessarily the case. The United States during the era of slavery, Stalin's Russia, Hitler's Germany, and Assad's Syria demonstrate that the "interest" of a political community can be immoral. If one counters, that by "interests" we mean only those interests that have a reasonable moral justification, then we are no longer discussing sovereignty, as we have abandoned its content neutrality. As such, we are led to a conclusion in which the assignment of the right of nonintervention to a political community depends on the moral nature of the interests being pursued. In which case, the scope of sovereignty cannot be accommodated, at least not fully.

In addition, Benn and Peters' argument rests on a similar assumption to one made by Walzer in which the state is comprised of a single people with a univocal collective interest. This is implied by the claim that a political community has interests that it knows and upon which it can act. This requires a level of uniformity inconsistent with the fact that most states are comprised of numerous peoples with often conflicting interests.

Alternatively, one might argue that the interests of the political community can be separated from the individual peoples of which the political community is comprised. In which case, an argument is needed to defend the proposition that the political community should be understood to be ontologically distinct from the peoples of which it is comprised, and that this distinction carries with it a separate and independent moral status for the political community in question. This proposition was discussed above and is untenable.

One might argue that the interest of the political community is tied, not to the interests of the peoples of which it is comprised, but is an aggregation of the interests of the individual members of the political community. Such interests, however, can't simply be the interests of the individual members. For one, the interests of a political community include matters that might not be appropriate to consider as a matter of individual interest. It is also the case that individual interests might be better protected by individual rights, thus making a collective right unnecessary.[33] Finally, it is even more unlikely that individual interests will have the necessary uniformity or coherence. In the end, there is little reason to believe that this argument can accommodate the scope of sovereignty.

Notwithstanding this, it is worth assessing whether Benn and Peters' argument can accommodate the content of state sovereignty. As has been argued, to serve as a moral justification for the right of nonintervention, the interests of the political community in question must be moral in nature. This does not support a form of sovereignty with the normative breadth of the notion of sovereignty under discussion here. In addition, since the argument is based on the proposition that the state knows and cares for the interests of the political community better than any outsider could,

[33]Wellman (1999).

it must be true that states are in an epistemically privileged position to know what the political community's interests are and how best to promote them. Thanks to the telecommunications revolution our world is much more epistemically interconnected, which makes this claim far from obvious. When intervention is backed by independent moral reasons, it is, at least, possible that outsiders have a better understanding of the interests of the relevant political community.

Whether a state knows what the interests of the political community of which it is comprised are, is an open question, as is the claim that the state cares about such interests more than any outsider could. As such, the argument cannot accommodate the nature of state sovereignty since, due to the existence of an open question, the argument does not accommodate state sovereignty as a second-order exclusionary reason.

4.2.2.4 Non-Analogical Arguments: Jovan Babic

The final non-analogical argument to consider is an argument from anti-paternalism made by Jovan Babic.[34] Babic's argument is cosmopolitan, and arguably reductive, in nature, based on the proposition that anti-paternalism in the international arena is essential to respect the choices of individuals. Elsewhere, I have noted that Babic argues that "[r]espect for individual human rights requires that states show tolerance for the conceptions of the good chosen by individuals in other states even if such choices appear to be morally or practically wrong."[35] Since there is no objective way to determine which conception of the good is best, a neutral principle is needed to avoid unjustifiable impositions of conceptions of the good on communities whose members did not choose them.[36] Therefore, Babic concludes, we need sovereignty to protect against such moral imperialism.[37]

One reason to be skeptical of the ability of Babic's argument to accommodate the scope of state sovereignty is one that has been discussed at length already. The right to choose the political good attaches to political communities. Political communities are not equivalent to states. Therefore, there is a conceptual gap between political communities and states that must be filled before the argument can accommodate states as rights-holders.

Second, the argument is about the ability of individuals within a political community to choose a conception of the good, including the political good, for themselves. As an argument grounded in human rights, one needs to explain why states should be attributed such rights over individuals. There are two interpretations of this argument, one with which I am sympathetic and the other I find unconvincing. The former takes the form of an instrumental argument, the sort of argument to be

[34] Babic (2003), pp. 52–54.

[35] Kassner (2012), p. 104.

[36] Babic (2003), pp. 52–54.

[37] *Ibid.*

discussed below. As to the latter, one might argue for the separate moral status of the state. For all the reasons I have already discussed, I find such arguments unconvincing.

In addition, if a state's claim to the right is dependent on a political community's choice of a conception of the political good, what if there is no single conception of the political good receiving overwhelming support? It would be unreasonable to require unanimity. But how much support is necessary? Can we even ask this question, nevertheless answer it, without imposing some conception of the political good, like fairness or equal treatment? In which case, if that is, in-and-of-itself, an affront to the human rights of individuals, then it is unclear what this means other than whatever a state chooses to do is implicitly a manifestation of the conception of the political good chosen by the members of their political community. The argument becomes nothing more than a tautology.

In addition, because, in nearly all circumstances, the state's choices and actions will reflect only one amongst a number of competing conceptions of the political good endorsed by the members of the relevant political community, why isn't respecting a state's "choice" also a failure to protect the choice of individuals who disagree? In the end, though the principle appears neutral on its face, in practical terms, it is often biased in favor of the powerful. In either case, it is hard to see how granting a state a right based on this argument can be justified.

As for the ability of Babic's argument to accommodate the content of state sovereignty, a commitment to neutrality over conceptions of the political good does not imply a right with the same normative breadth as the right of nonintervention. In addition, if the basis for the right is epistemic uncertainty over which conceptions of the good are correct, this is not the same as saying that we have ***no ability*** to distinguish between better and worse conceptions. Unreasonable conceptions of the good can be rejected because in making that determination we are not invoking some controversial notion of the good, but merely requiring that a conception not be unreasonable. For example, a conception of the good grounded in the proposition that ***those*** people who don't look like me aren't "really human" can be rejected as unreasonable without being unjustified moral imperialism. As such, the argument cannot accommodate the content of sovereignty.

Another problem with the ability of Babic's argument to accommodate the content of sovereignty derives from the fact that it is often true that the ability of individuals to choose their conception of the good will conflict with the political community's choice. However, the value of the political community's choice is dependent on respecting the human rights of individual members of the community. When such values conflict, it is unclear why a political community's choice should be given priority. In which case we are, yet again, faced with an open question. As such, not only is this argument incapable of accommodating the content of state sovereignty, it cannot accommodate the nature of sovereignty as a second-order exclusionary reason.

4.2.2.5 Instrumental Arguments

The last set of arguments are instrumental in nature and are based on the role state sovereignty plays in the achievement and/or promotion of valuable moral objectives sought by the international community. Namely, state sovereignty is often argued to be justified by its contribution to international peace and security[38] and political self-determination.[39] These are not mutually exclusive bases for the argument, as such I will consider them as distinct aspects of a single argument.

Looking first at the potential justification to be found in the relationship between sovereignty and international peace and security, as Mark W. Janis noted, "[t]he conflicting allegiances of Europe had contributed to the terrible toll of confusion, death, and destruction from 1618 to 1648. In the mid-seventeenth century many Europeans sought a simpler, and it was hoped, safer set of loyalties."[40] It was the nascent idea of sovereignty found in the Peace of Westphalia that enabled the emergence of (a semblance of) international order.[41]

As stated, however, this historical understanding does not provide a moral justification for sovereignty. Instead we have a conventional understanding of a norm that requires a justification. Why should we think that sovereignty serves international peace and security? In *Perpetual Peace*, Immanuel Kant rests his argument on the proposition that peace between political communities is a desirable moral end, arguably the most important end within the international arena, not necessarily for its own sake, but for what such peace would mean to the freedom and autonomy of individuals.[42] Essential to achieving peace between states, according to Kant, was the imperative that "[n]o state shall forcibly interfere in the constitution and government of another state."[43] Thus, each state has a right of nonintervention to protect against the threat of violence and disorder in the international arena.

As to the relationship between political self-determination and sovereignty, as was noted above, Hans Morgenthau contends that during the French Revolution the French argued that sovereignty promotes and protects political self-determination.[44] Recognizing that a historical relationship between sovereignty and political self-determination exists does not provide a moral justification. Since the argument is an instrumental one in which the justification is grounded in the moral value of political self-determination, to understand the justification we must understand both the value of political self-determination and the relationship between sovereignty and political self-determination.

---

[38]Kassner (2012), pp. 26, 104.

[39]*Ibid.*, p. 104.

[40]Janis (1999), p. 159.

[41]Smith, p. 204.

[42]Kant (2003).

[43]Kant.

[44]Morgenthau.

Regarding our understanding of the value of political self-determination, A. John Simmons has offered one of the most compelling arguments that both serves as a justification for, as well as an explanation of, the value of political self-determination. Simmons argues that due to the pervasive impact of law and public policy on our lives, individual self-determination means little without political self-determination. And, as Stanley Hoffman has argued, "in a world of states, sovereignty protects one against outsiders trying to topple the government or to set up a puppet regime or to impose their views of what is good and right – hence the particularly strong attachment of countries recently liberated from colonial rule to the principle of nonintervention."[45]

Can such instrumental arguments accommodate the constitutive elements of state sovereignty? Beginning with the scope of sovereignty, understanding the answer depends on the way in which any instrumental account can justify the attribution of rights at all, since, for many, an instrumental defense of rights is, at the very least, counterintuitive because rights tend to constrain—rather than depend on—instrumental reasoning.

In explaining how an instrumental account of value could justify the ascription of rights, the argument relied upon here is based on L. Wayne Sumner's understanding of the moral foundation of rights.[46] Sumner contends that the attribution of rights can be justified by the role rights-based reasoning plays in the pursuit of other moral goals. For example, you might attribute rights to free speech to individuals, not because free expression is inherently valuable, but because free speech is the best available ***means*** to truth and knowledge. As such it is necessary to protect free speech from simple cost-benefit analysis and employing rights-based reasoning provides the required protection.

The argument under consideration relies on two propositions. First, it must be the case that international order, and political self-determination are worthwhile moral objectives of the international community. And, second, that political self-determination and international order are advanced by rights-based reasoning grounded in sovereignty. Regarding the former, to the extent that we understand war and interstate violence to be morally bad and worth preventing, then it is reasonable to assume that such prevention is a morally worthy goal of the international community. As to political self-determination, if we believe that, as a moral matter, people should have the ability to decide for themselves how their lives will be governed, then political self-determination is likewise a morally worthy goal to be pursued and protected by the international community.

But, is sovereignty an effective means of advancing either international order, or political self-determination? Let us first consider what sovereignty demands. Taking the right of nonintervention as the content of sovereignty, one need only consider the implications of a normative system in which this right is respected to understand how it is instrumentally valuable to both international order and political

[45]Hoffman, p. 19.

[46]Sumner.

self-determination. As to the former, if states refrain from violating the political independence and territorial integrity of other states, the risk of interstate conflict is greatly reduced, and international order is promoted. Similarly, the ability to be self-governing, and by implication politically self-determining, is promoted by adherence to the demand that states refrain from interfering in one another's political affairs. One might even argue that the instrumental relationship between sovereignty and political self-determination is a constitutive one, that to be politically self-determining a political community must be protected from outside interference.

In addition to these more abstract reasons for understanding the instrumental relationships between sovereignty and the various valuable objectives it is thought to serve, there is ample empirical evidence supportive of the instrumental relationship between state sovereignty and political self-determination and international order. Respect for sovereignty has changed the nature of international relations, limiting interstate conflict and protecting former colonies from outside interference.

One might object, however, that this is merely the epiphenomena of realist politics. That sovereignty is a convenient excuse to refrain from becoming entangled in other people's problems, and that sovereignty is disregarded whenever it suits a great power. In support of the former, one might cite the use of Rwandan sovereignty in the Spring of 1994 as an excuse for inaction by the international community during the Rwandan genocide. As to the latter, one need only consider the 2014 invasion of Crimea by the Russians as evidence of the willingness of global powers to disregard sovereignty when it suits them.

The objection is based on an assumption about the relationship between adherence to a norm and the existence of the norm, that the existence of a norm is tied to obedience to the norm. How much obedience is required? It cannot be perfect adherence to a norm, because in the real world such perfect adherence is practically unattainable. If, however, it is something less than complete obedience that is required, what is enough? I contend that the standard should be tied to what is necessary for the norm to be recognized as one that ought to be followed. In other words, does it serve as a viable source of criticism and defense within a normative system? This does not require perfect adherence, but it does require that those subject to the norm recognize it as an independent source of criticism or defense.

Taking this as an acceptable means of distinguishing between "real" norms and those that only portend to be, neither of the examples supports the proposition that state sovereignty is merely a pretext for realist considerations. The choice by the international community not to intervene in Rwanda has been recognized as a failure; and, at least in part, led to the development of the Responsibility to Protect doctrine which provides a more nuanced understanding of the responsibilities of sovereign states and the international community.

As to the Russian invasion and annexation of Crimea, this has not been met with muted acceptance. Rather, it has resulted in outrage and sanctions with much of the condemnation tied to the consensus around the claim that Russia's actions are a violation of Ukrainian sovereignty. The fact that the international community has not chosen to use military force to repel the Russian forces does not mean that sovereignty is being ignored. In both cases, the violation of sovereignty or the inappropriate use of it as an excuse to avoid action have been met with criticism and sanctions in which sovereignty has played a significant role in the critique itself.

Having addressed this possible objection, the discussion can now return to assessing whether the instrumental argument can accommodate the scope of state sovereignty. If it is true that political self-determination and international order are served by attributing sovereign rights to states, then the argument can accommodate the scope of sovereignty. To a large extent, the ability of the instrumental argument to accommodate the content and nature of sovereignty has already been discussed, but I will provide a brief recap to ensure against any confusion. The argument must be able to accommodate the right of nonintervention. If we take political self-determination and international order to be worthy moral goals of the international community—a proposition that I have already defended—and we understand the right of nonintervention as an effective means of protecting and promoting these goals, then the right of nonintervention is justified by its instrumental relationship to such goals.

Regarding the ability of the argument to accommodate the nature of sovereignty, the answer is found in the fact that what is being justified is a right and the rights-based reasoning it engenders. The right of nonintervention is, by its nature, a second-order exclusionary reason. As such, since the instrumental argument can justify the right of nonintervention, it can accommodate the nature of sovereignty as well.

Before concluding, since one of the critiques levied against the previous arguments was that they often led to open questions about the moral underpinnings of the arguments in question, one might argue that the instrumental nature of the justification leads to a similar concern. One could even interpret my previous analyses as leading to the proposition that most of the prior arguments, in their most palatable forms, are instrumental. State sovereignty, under such arguments, is supported by its ***presumed*** instrumental relationship to the goods sought.

To understand the "open-question" critique and why it does not undermine the instrumental justification for sovereignty, one needs to understand presumptions and how they are justified. When we are presented with a question of how we ought to act, presumptions often take the form of "rules of thumb". We act ***as if*** there is a rule governing our behavior because we presume that the correct course of action is ***more likely*** to come about by following the rule. Presumptions are provisional rules, shifting the burden of argument to those challenging the presumed rule or course of action. Presumptions are not based on conceptual certainties, rather they are dependent on probabilities. As such, justifying a presumption does not require certainty, rather it requires that we assess what the evidence supports.

What undermines the arguments discussed above, leading to my open-question critique, is that there is little evidence to support the claimed presumptive instrumental relationships between sovereignty and the various moral goods claimed to be promoted by respecting sovereignty. The same cannot be said when it comes to the relationship between sovereignty and either, political self-determination or international order. Such evidence can be mustered, and the presumption justified. Consequently, the instrumental justification for state sovereignty discussed above can be justified as well.

4.3 Conclusion: Limits on the Legitimate Zone of Sovereign Discretion Implied by an Instrumental Justification for State Sovereignty

Having defended the instrumental argument as the best available moral justification for sovereignty, it remains to be determined what this implies for the legitimate zone of sovereign discretion. One might conclude that this implies a zone of sovereign discretion with limited normative breadth, that the argument is exhausted by the instrumental relationship between state sovereignty and international order and political self-determination. This would fail to understand the nature of the instrumental argument. It is merely a contingent matter that the argument is limited to international order and political self-determination. Any matter of moral concern facing the international community ***could*** serve as the basis for the extension of the legitimate zone of sovereign discretion.

To be more specific, if extending sovereign discretion to include a moral challenge facing the international community is the best means to responding to the challenge, then there is an instrumental reason for doing so. If we look to human rights as an example, assuming that the provision and protection of human rights is a moral objective that ought to be pursued by the international community, under the instrumental argument, if extending the legitimate zone of sovereign discretion to include human rights is one of the most effective means of providing and protecting human rights, then there is an instrumental argument for doing so.[47]

Finally, what does this argument imply for our understanding of the limits to the legitimate zone of sovereign discretion? Whether a particular moral challenge facing the international community ought to be included within the legitimate zone of sovereign discretion depends on whether there is evidence or a viable argument to be made that responding to the challenge will be aided by so extending sovereign discretion.

References

Babic, Jovan. 2003. Foreign Armed Intervention: Between Justified Aid and Illegal Violence. In *Humanitarian Intervention: Moral and Philosophical Issues*, ed. Aleksandar Jokic, 45–70. Peterborough, ON: Broadview Press.

Beitz, Charles R. 1999. *Political Theory and International Relations*. Princeton, NJ: Princeton University Press.

Benn, S.I., and R.I. Peters. 1965. *The Principles of Political Thought: Social Principles and the Democratic State*. New York: Free Press.

Hall, William E. 1924. *International Law*. 8th ed, 347. Oxford: Clarendon Press.

Hoffman, Stanley. 1996. *The Ethics and Politics of Humanitarian Intervention*, 19. Notre Dame, IN: University of Notre Dame Press.

---

[47] See Shue (2004).

Janis, Mark W. 1999. *An Introduction to International Law*. 3rd ed. New York: Aspen Law & Business.
Kant, Immanuel. 2003. *Perpetual Peace*. Indianapolis, IN: Hackett Publishing.
Kassner, Joshua J. 2012. *Rwanda and the Moral Obligation of Humanitarian Intervention*. Edinburgh, UK: Edinburgh University Press.
Margalit, Avishai, and Joseph Raz. September 1990. National Self-Determination. *The Journal of Philosophy* 87 (9): 439–461.
Mill, John Stuart. 1867. A Few Words on Nonintervention. In *Dissertations and Discussions*, vol. III. London: Longmans, Green, Reader, and Dyer.
———. 1991. *On Liberty (1859) from On Liberty and Other Essays*, 14. Oxford: Oxford University Press.
Ohlin, Jens David. 2015. *The Assault on International Law*. Oxford: Oxford University Press.
Rawls, John. 1999. *The Law of Peoples*. Cambridge, MA: Harvard University Press.
Shue, Henry. 2004. *"Limiting Sovereignty," from, Jennifer M. Welsh, Humanitarian Intervention and International Relations*. Oxford: Oxford University Press.
Walzer, Michael. Spring 1980. The Moral Standing of States. *Philosophy & Public Affairs* 9 (3): 209–229.
———. 2000. *Just and Unjust Wars: A Moral Argument with Historical Illustrations*. 3rd ed, 87. New York: Basic Books.
Wellman, Carl. 1999. *The Proliferation of Rights: Moral Progress or Empty Rhetoric*. New York: Westview Press.
Wellman, Christopher Heath. Spring, 1995. A Defense of Secession and Political Self-Determination. *Philosophy & Public Affairs* 24 (2): 142–171.

Chapter 5
Introducing a Framework for Assessing the Moral Limits of Sovereignty and Sovereign Discretion

In Chaps. 3 and 4, two different approaches to understanding the moral limits to the legitimate zone of sovereign discretion were identified and explicated, those inherent to the notion of sovereignty itself and those implied by an instrumental moral justification for sovereignty. These limitations are the grounding for a framework to be used to assess the moral limits to the legitimate zone of sovereign discretion and, by implication, the moral limits of state sovereignty.

One may wonder about the practical significance of these limitations and the framework being proposed. How does the understanding of the limits to the legitimate zone of sovereign discretion implied by these arguments differ, if at all, from the current understanding? And, what, if anything, do such moral limits tell us about the role sovereignty ***ought*** to play in the normative framework of international relations? This chapter is both an answer to these questions and serves as the first step in the transition from theory to practice.

5.1 Current Understanding

The first step in understanding the practical significance of the moral limits to sovereign discretion implied by the limitations identified in Chaps. 3 and 4, is to understand the role sovereignty plays in the existing normative framework of international relations. Specifically, one needs to understand the zone of sovereign discretion states are taken to have and the role sovereignty is taken to play in the existing normative framework of international relations before one can understand the difference that the limitations identified in Chaps. 3 and 4 make.

As such, this discussion proceeds from the account of the conventional notion of sovereignty provided in Chap. 2. That account of sovereignty serves as the basis for our understanding of the zone of sovereign discretion this conventional notion of sovereignty is taken to imply. The implications from that conventional

J. J. Kassner, *Climate Change and Sovereignty*, AMINTAPHIL: The Philosophical Foundations of Law and Justice 10, https://doi.org/10.1007/978-3-030-73578-4_5

understanding of the zone of sovereign discretion explains the role sovereignty plays in the existing normative framework of international relations.

With these goals in mind, it is important to note that sovereignty, even under the conventional understanding, though expansive, is not absolute. As such, we must identify the underlying commitments that limit the zone of sovereign discretion within the existing normative framework of international relations. It is widely accepted that sovereign discretion is limited by the international community's commitments to international order and political self-determination.

It is not my intention to be critical of these commitments or the understanding of the limits to sovereign discretion they are taken to imply. Rather, I am simply providing an explanation of that understanding. Such critical reflections have already been discussed at length in the preceding chapters and will also be addressed in the chapters that follow when the discussion shifts to an assessment of the moral limits to the legitimate zone of sovereign discretion.

With that said, the international community's commitments to international order and political self-determination have been taken to imply a prohibition on actions that would interfere with the territorial integrity or political independence of any sovereign state. As such, under the existing normative framework of international relations, those commitments imply that actions that interfere with the territorial integrity or political independence of sovereign states are taken to be beyond a state's sovereign discretion.

There is also a growing consensus that human rights are a limitation on sovereign discretion. The exact nature of this limitation has yet to be fully fleshed out. For example, RtoP is proffered as an understanding of the relationship between sovereignty and human rights, wherein claims to sovereignty are dependent on respect for human rights. However, the underlying theory, impotence of the structural reforms accompanying the doctrine and actual practice make any claims that human rights are, as a matter of conventional fact, limitations on sovereign discretion uncertain at best.

Nonetheless, even if human rights are included as a limit on sovereign discretion, the limitation is a narrow one, leaving many matters within the zone of sovereign discretion. This broad conventional understanding of sovereign discretion has resulted in the unfounded assumption identified earlier, in which states have discretion over whether and how to respond to new and emerging challenges facing the international community, including the challenge of climate change.

An example may help to further illuminate both the conceptual and practical significance of this presumption. Recently, the international community, through the Paris Climate Agreement (PCA), came together and, for a brief moment, seemed to be taking on the challenge of climate change. But, certain aspects of the agreement assumed that whether and how a state should react to the challenge of climate change was a matter within a state's sovereign discretion. The agreement is not legally binding, and action in accordance with the agreement is voluntary. Even the emissions commitments are to be “Nationally Determined”.[1] Agreements, like the PCA, are of tremendous rhetorical value, but that does not negate the fact that whether and how a state acts, even under the PCA, remains within the state’s discretion.

5.2 Revised Understanding: Legitimate Zone of Sovereign Discretion

With this understanding of the zone of sovereign discretion found in the conventional understanding of the normative framework of international relations in hand, we should turn to a brief review of the limits to the legitimate zone of sovereign discretion discussed in Chaps. 3 and 4. This brief overview provides the basis for a unified account of the limits to the legitimate zone of sovereign discretion, which is then, at various points in this discussion, juxtaposed with the conventional understanding of the limits to sovereign discretion to highlight the conceptual and practical differences between the two.

The review is focused on three elements. First, what are the key normative aspects to the limits to the legitimate zone of sovereign discretion in the unified account. Specifically, what are those aspects that are necessary for understanding the practical significance of the difference between the existing understanding of the limits to sovereign discretion and the unified account? Second, what does the unified account imply for how we ought to evaluate new and emerging challenges facing the international community? Finally, what implications does the unified account have for the role sovereignty ***ought*** to play in the normative framework of international relations?

5.2.1 Inherent to the Notion of Sovereignty Itself

First, understanding the key normative aspects of the limits to the legitimate zone of sovereign discretion inherent to the notion of sovereignty itself, requires that we revisit the underlying argument. Recall, the argument is based on the proposition that sovereignty is comprised of various rights held by, and correlative duties borne by, states, namely, the rights to territorial integrity, political independence, and sovereign equality and duties owed to other members of the international moral community. Collectively, these rights are taken to imply a principle of nonintervention, including a right of nonintervention and a correlative duty not to interfere with the territorial integrity or political independence of other sovereign states.

---

[1]Conference of the Parties, Adoption of the Paris Agreement U.N. Doc. FCCC/CP/2015/L.9/Rev/1 (Dec. 12, 2015) (“Paris Climate Agreement”).

From this we can conclude that one key normative aspect of our understanding of the limits to the legitimate zone of sovereign discretion is that, so long as a state is acting within the normative breadth of its rights and fulfilling the duties it owes to others, it is free to act in ways that others may find morally or practically objectionable. And, other states and the international community are duty-bound to refrain from interfering with actions taken by a right-holding state that are within the normative breadth of a rights-holding state's rights—the state's zone of sovereign discretion.

At first glance, this may lead some to believe that this implies an understanding of sovereignty with the same discretion as that found in the extant normative framework of international relations. This, however, fails to take into account that our effort to understand the limits inherent to the notion of sovereignty itself is tied to the philosophical nature of the inquiry. It is an effort in conceptual analysis, wherein the validity of the norm is assumed and then an assessment is made regarding what we take the conventional understanding to imply. Understanding the extant conventional understanding of the zone of discretion in the normative framework of international relations is about what, as a matter of empirical observation, the zone of sovereign discretion ***is taken to be.*** It is not a product of philosophical or conceptual analysis.

What does this mean for how we understand the legitimate zone of sovereign discretion? How do we determine whether the new and emerging challenges facing the international community are within the legitimate zone of sovereign discretion? The limits to the legitimate zone of sovereign discretion inherent to the notion of sovereignty itself, imply that actions that violate another state's rights or that fail to fulfil the duties borne by a state are outside the legitimate zone of sovereign discretion. Thus, a state's discretion extends to actions that do not violate another state's rights to territorial integrity, political independence, sovereign equality, or non-intervention, and to actions that do not involve a failure to fulfill other duties that the duty-bearing state owes to others.

As noted, we are concerned with the deliberative processes by which reasons for action are considered and through which an agent determines what they ought to do based on such reasons and reasoning. Under this understanding, sovereignty operates both as a first-order reason for action and as a second-order exclusionary reason. The first-order aspect of sovereignty is relatively familiar. As a norm it provides reasons directly for or against particular actions.

Recall, second-order exclusionary reasons, in addition to being first-order reasons for action, exclude from consideration countervailing first-order reasons for action that would otherwise, but for the exclusionary impact of the second-order reason, be relevant. As an example, imagine that you and I agree to meet for lunch on Friday; however, on Thursday you are invited to attend a concert that conflicts with our lunch date. In terms of what would bring you more pleasure, the concert is—as one might expect if they know me—the overwhelming winner. If this were merely a conflict of first-order reasons, then you are arguably justified in standing me up for our lunch date. And, when I asked where you were, you could simply recount the fact that the concert brought more pleasure and that should serve as sufficient justification.

Our agreement, however, has an additional impact on our practical deliberations. It is not merely a first-order reason for you to meet me for lunch, nor is it the case that simple first-order instrumental reasoning remains unaffected. Rather, the agreement creates a duty that we owe each other, and that duty is a second-order exclusionary reason. You shouldn't consider the concert as an option, at least not without seeking a release from our agreement.

Second-order exclusionary reasons are not, however, absolute. Their impact on our practical deliberations are limited in two ways. Not all first-order reasons are excluded. There are some first-order reasons that remain outside the reach of the exclusionary effect of the second-order reason in question. In addition, second-order reasons can be outweighed by other second-order reasons. To explain, let us return to the example just given. Imagine that you are a voluntary firefighter and at the time of our lunch date you are on call. Just minutes before we are scheduled to meet for lunch, the local alarm sounds, and you are called to a fire. In that case, you would be justified in standing me up because your duty as a firefighter is a second-order exclusionary reason; and, in this case, it (at least arguably) outweighs your obligation to meet me.

Returning to our discussion of sovereignty as a second-order exclusionary reason, sovereignty excludes from consideration those reasons for action the fulfilment of which would require a state to interfere with the rights of another state to their political independence or territorial integrity. In other words, many reasons for action that would, without sovereignty, be relevant to determining what a state ought to do are rendered irrelevant by the exclusionary impact of sovereignty as a second-order exclusionary reason.

In instances where a state's actions violate another state's rights to territorial integrity, political independence, sovereignty equality, or non-intervention, the duty of the interfering state to refrain from engaging in such actions gives rise to a weighty first order reason to stop. In addition, to the extent that the interfering state refuses to stop, sovereignty implies a weighty reason, and ***a*** moral reason, for the international community to act to stop the interfering state.

As an example of what this might look like in the practical deliberations of a state, imagine that democratic legitimacy is ***the correct*** account of political legitimacy, and without a democratically legitimate government, a state's claim to authority is suspect. Also imagine that a state has chosen to pursue a model of benevolent authoritarianism instead and that implementing democratic political processes would require the international community to violate the authoritarian state's political independence. That such authoritarian rule undermines the state's claim to authority is precluded from consideration to the extent that action seeking the implementation of democratic political processes would require violation of the subject state's sovereignty.

5.2.2 *As Implied by Best Available Moral Justification for State Sovereignty*

In Chap. 4 it was argued that the best available moral justification for sovereignty is an instrumental one, wherein sovereignty is justified by its role in achieving other valuable moral ends sought by the international community. Understanding the limits to the legitimate zone of sovereign discretion implied by its instrumental value, requires a clarification of the instrumental justification itself. As noted, in the existing normative framework of international relations, the valuable moral objectives pursued by the international community include international order and political self-determination. One might conclude from this fact of our shared political history, that the instrumental justification for sovereignty is therefore limited to the role it plays in the pursuit of these valuable moral objectives.

This evinces a failure to understand the nature of the justification. For such a narrow understanding of the justification's implications to be correct, it would have to be the case that the valuable moral objectives of the international community are limited to international order and political self-determination. However, the fact that sovereignty is an artefact of our creation implies that the justification for sovereignty is constructed on the proposition that respecting state sovereignty enhances the international community's ability to respond the challenges it faces. There is nothing inherent to this underlying justification that limits it to the existing objectives sought by the international community. Nor, is there any reason to conclude that these are the only matters that could be served by sovereignty. Rather, we are left with an understanding that the value of sovereignty, in relation to any morally valuable goal sought by the international community, is tied to the role sovereignty plays, or could play, in the pursuit of the goal.

Thus, extending sovereign discretion based on the instrumental value of sovereignty is contingent and malleable. It might also be the case that sovereignty may, one day, prove to be unjustified *in toto* if it no longer proves useful to the international community's efforts to respond to any of the challenges it faces. For our purposes then, the legitimate zone of sovereign discretion depends on the instrumental relationship between sovereignty and the new and emerging challenges facing the international community.

With this understanding we can now directly assess what this implies for our understanding of the limits to the legitimate zone of sovereign discretion. The legitimate zone of sovereign discretion should extend to those challenges facing the international community the resolution of which would be aided by including the challenge within the legitimate zone of sovereign discretion.

To some, the argument may seem circular. The moral justification, however, does not start with the norm or principle that is being justified. Rather, the justification begins with the moral objective serving as the ultimate source of the moral justification—the thing that we are trying to achieve. Consequently, the justification depends on the extent to which the norm or principle for which we are seeking a justification is a means to our ultimate goal. Much of the moral work is done in defending the claim that the objective is morally valuable.

An additional implication derived from this understanding of the limits to the legitimate zone of sovereign discretion relates to those matters in which sovereignty is not a path to resolution or protection. In short, under an instrumental justification of the sort we are discussing here, to be justified, including the matter within the legitimate zone of sovereign discretion can't undermine our efforts to achieve the goal being sought. If it does, then the norm cannot be justified, at least in relation to the matter under consideration.

This does not, however, settle the matter. Our understanding remains incomplete. The fact that sovereignty is instrumentally valuable to the international community's efforts to respond to a challenge is not enough to justify the attribution of sovereign discretion over the matter at hand. Rather, this fact merely opens a broader inquiry into the comparative value of sovereignty as a means as compared to other available means. In determining which path ought to be pursued a wide array of ethical and practical considerations must be considered in order to determine what path to choose.

Returning to the example of human rights, if we assume that the provision and protection of human rights are valuable moral objectives that the international community is (or ought to be) pursuing, then what is instrumentally justified is determined by what actions, institutions, norms, . . . , etc. are effective means to the provision and protection of human rights. It is unlikely that there is a single path to this objective. Not all paths will be equally effective or ethically acceptable.

In an ideal world we might be able to guarantee everyone's rights through universal recognition of the rights people have and universal fulfillment of the correlative duties borne by others. This would be an effective model if there were no disagreements over and no defections from the moral order. In the real world, however, there are disagreements and defections. Consequently, there is a need for social and political institutions to ensure the fulfillment of the moral demands implied by human rights. Which institutions we should choose depends on comparative effectiveness and ethical acceptability.

In addition, without more, it is not necessarily the case that the fact that the international community's efforts to respond to a challenge are not served by including the challenge within the legitimate zone of sovereign discretion means that sovereign discretion ought not be extended to include the challenge. The instrumental value of sovereignty needs to be understood holistically. If, for example, excluding human rights from the legitimate zone of sovereign discretion would undermine international peace and security, then we would need to engage in an independent assessment of the relative value of the two ends being sought and the impact of either including human rights within versus excluding human rights from the legitimate zone of sovereign discretion would have on the international community's ability to respond to any of the affected ends.

With that said, there are three lessons we can take from our understanding of the limits to the legitimate zone of sovereign discretion implied by the instrumental justification for sovereignty. First, the zone of sovereign discretion extends only as far as its instrumental value. Second, the fact that sovereignty is instrumentally valuable is not enough to justify the inclusion of a matter within the legitimate zone of sovereign discretion. Finally, we need to engage in a more holistic evaluation, comparing sovereignty with other possible paths to resolution, taking both practical and ethical considerations into account, and accounting for the impact excluding the matter from the legitimate zone of sovereign discretion might have on other moral objectives sought by the international community.

Our attention should now turn to identifying the limits to the legitimate zone of sovereign discretion implied by the instrumental justification for sovereignty and assessing the implications this has for the role sovereignty should play in the normative framework of international relations. Recall, it is the exclusionary aspect of sovereignty that defines the legitimate zone of sovereign discretion and that must be justified for any extension of the zone of sovereign discretion to be warranted. Whether the legitimate zone of sovereign discretion ought to be extended depends on whether the international community's efforts to respond to a challenge or achieve a valuable moral end are aided by extending sovereign discretion over the matter in question. If not, or if there is a more effective and ethically preferable means available, then the matter should be considered outside the legitimate zone of sovereign discretion.

It is worth noting, that even if a matter is excluded from the legitimate zone of sovereign discretion under the instrumental argument, this would not mean that the international community is justified in intervening. It would merely imply that the international community's response is not governed by sovereignty. For example, let us assume that the use of the death penalty in the United States is a violation of basic human rights and that this justifies piercing the veil of sovereignty, providing a powerful moral reason in favor of interference with the internal political affairs of the United States.

This would not settle the matter, as it is but one reason among many that needs to be considered. The harm caused by such interference, especially if a military intervention would be required to affect the goal, would, all-things-considered, greatly outweigh the benefit, even if achieved. Importantly, the practical deliberations leading to this conclusion are based on an independent moral evaluation of the reasons for and against intervention, not influenced by sovereignty.

Further, an instrumental account of the moral justification for sovereignty and the limits to the legitimate zone of sovereign discretion does not imply an inviolable boundary around the legitimate zone of sovereign discretion. Rather, what challenges fall within the legitimate zone of sovereign discretion depends on an application of the three elements discussed previously. If extending the zone of sovereign discretion to include a matter is the best available means (both practically and ethically) to responding to a challenge, then the legitimate zone of sovereignty ought to be so extended.

There may also be an argument for extending the zone of sovereign discretion in a particular case, if failing to do so will greatly undermine the ability of the international community to achieve other morally valuable objectives. In either case, the practical deliberations of states over how to respond to the matter in question, will be impacted by the second-order exclusionary nature of sovereignty. If, however, such an extension of sovereign discretion is either not the best available means, or would, in fact, undermine the international community's efforts to respond to a challenge, then sovereign discretion ought not be extended; in which case, the practical deliberations of states would not be fettered by sovereignty.

In terms of assessing the difference between this understanding of the legitimate zone of sovereign discretion and the extant zone of sovereign discretion found within the current conventional understanding of sovereignty and sovereign discretion, a general comparison is not possible. Recall, that the limits to the legitimate zone of sovereign discretion implied by the instrumental justification can only be discovered in a piecemeal fashion; consequently, the comparison between the conventional understanding of sovereignty and sovereign discretion and that implied by the instrumental justification must, likewise, proceed in a piecemeal, challenge by challenge, fashion.

5.2.3 A Unified Account of the Legitimate Zone of Sovereign Discretion

Neither account of the limits to the legitimate zone of sovereign discretion is ***independently*** sufficient as an understanding of the legitimate zone of sovereign discretion. Rather, a unified account is needed. Only when we have such an account can we adequately understand the limits to the legitimate zone of sovereign discretion. In seeking such an account, there are at least two paths that could be pursued. One would involve a conceptual reconciliation between the two understandings, identifying and explicating their logical relationship. The other would seek a practical unification, providing a workable framework for addressing new and emerging challenges facing the international community.

For both methodological and pragmatic reasons, in the discussion that follows a practical unification is provided. As a methodological matter, the instrumental justification discussed in Chap. 4 already provides an adequate explanation of the conceptual relationship between the instrumental justification and the notion of sovereignty at issue. In addition, though a thorough conceptual reconciliation would explore the logical relationships between the limits to the legitimate zone of sovereign discretion implied by both accounts, the primary purpose of this book is practical in nature and exploring those logical relationships would take us too far afield.

As such, and as to the pragmatic reasons for pursuing a practical unification, the objective here is to understand the implications these limits to the legitimate zone of sovereign discretion have for the practical deliberations of states and the international community and to further our understanding of the moral nature and limits of state sovereignty. A thorough conceptual reconciliation is unnecessary for either task.

As to the nature of the practical unification being proposed, it takes the form of an integrated framework for assessing whether a challenge facing the international community should be included within the legitimate zone of sovereign discretion. Constructing this integrated framework requires that we first understand how the two limits relate to one another. To be specific, in considering whether a challenge facing the international community ought to be included within the legitimate zone of sovereign discretion, we must explain how the two limits identified inform our answer.

The nature of the requisite explanation depends on whether we are making the determination for an ideal or a non-ideal world. If the former, it arguably makes sense to dispense with the limits inherent to the notion of sovereignty itself, as this understanding of the limits to the legitimate zone of sovereign discretion supervenes on the instrumental moral justification for sovereignty. The former understanding of the limits to the legitimate zone of sovereign discretion is (arguably) incapable of identifying a zone of sovereign discretion that is not already included within that implied by the underlying moral justification for sovereignty.

This, however, is not a project in ideal theory. As such and taking the non-ideal nature of the international arena into account, if the challenge under consideration is to fall within the legitimate zone of sovereign discretion, inclusion must be implied by both accounts. As to why, either understanding of the limits to the legitimate zone of sovereign discretion can serve as the basis for rejecting the proposition that a challenge falls within the legitimate zone of sovereign discretion, thus both limits must be addressed.

As both limits need to be included, one might wonder ***how*** exactly the two limits will be integrated. It makes the most sense to start with an application of the limits inherent to the notion of sovereignty itself because the history of inquiry, interpretation and application surrounding sovereignty provides us with an understanding of the content of the norm that is absent from the more speculative understanding of the instrumental value of sovereignty.

The fact that the unified account prioritizes the application of the limits inherent to the notion of sovereignty over those implied by the instrumental moral justification for sovereignty should not be taken as a statement about the relative importance of either. As noted, both limits stand as reasonable bases for objecting to the extension of the legitimate zone of sovereign discretion, and there will be challenges that according to the limits inherent to the notion of sovereignty are outside of the legitimate zone of sovereign discretion, but that, under the instrumental justification, ought to be included within the legitimate zone of sovereign discretion. It may also be the case that there is controversy over whether the matter falls within the zone of sovereign discretion inherent to the notion of sovereignty, in which case determining whether there is an instrumental justification for including the matter within the legitimate zone of sovereign discretion becomes determinative. The reason to prioritize the limits inherent to the notion of sovereignty is merely a methodological choice driven, in part, by the non-ideal nature of this project.

5.3 Significant Differences

There are significant differences between the extant conventional understanding of the zone of sovereign discretion and the legitimate zone of sovereign discretion implied by the integrated framework. As a first step to understanding the importance of the discussion thus far, a brief explanation of those differences, both theoretical and practical, is in order.

As to the former, "theoretical" does not mean that the difference to be discussed is not related to the practical orientation of the inquiry. Rather, the difference at issue is tied to the way in which each understanding of the limits to the legitimate zone of sovereign discretion determines whether a challenge facing the international community is properly thought of as falling within the legitimate zone of sovereign discretion. The most significant theoretical difference revolves around the assumption, underlying the conventional understanding, that all challenges facing the international community fall within the legitimate zone of sovereign discretion.

There is no such assumption under the integrated framework. As was argued, there has been little effort to justify this assumption, and the integrated framework is based on my demonstration that the assumption ***can't*** be justified. Under the integrated framework, whether a challenge falls within the legitimate zone of sovereign discretion is an open question. As such, we are not justified in saying whether any particular challenge falls within the legitimate zone of sovereign discretion until after applying the integrated framework.

Understanding the practical differences between the zone of sovereign discretion found in the extant conventional understanding and the legitimate zone of sovereign discretion implied by the integrated framework is informed by the theoretical difference discussed above. When considered in light of the fact that the challenges facing the international community have changed in fundamental and drastic ways, such theoretical differences have a significant impact on which, if any, of the new and emerging challenges fall (or ought to fall) within the legitimate zone of sovereign discretion.

5.4 Conclusion: A Framework for Assessing the Moral Limits of Sovereign Discretion

From this discussion we can conclude that the integrated framework is comprised of a lexical ordering of the two limits to the legitimate zone of sovereign discretion. The ordering is based on the non-ideal nature of the international arena and is informed by a preference for an ordering that is more practically useful when integrated in the normative framework of international relations. The limits inherent to the notion of sovereignty itself are lexically prior to those implied by the instrumental justification for sovereignty. Finally, both limits must be considered in assessing whether a challenge facing the international community falls (or ought to fall) within the legitimate zone of sovereign discretion.

From this understanding of the integrated framework, we can construct a general understanding of ***how*** the framework will be applied. To be specific, when confronted with a new or emerging challenge we begin by assessing whether the challenge in question falls within the legitimate zone of sovereign discretion inherent to the notion of sovereignty itself. We then assess whether the challenge falls within the zone of sovereign discretion implied by the instrumental justification for state sovereignty. If the challenge falls within the legitimate zone of sovereign discretion under both limits, then how the international community ***ought*** to respond to the challenge is limited by sovereignty and sovereign discretion.

If, on the other hand, neither understanding of the limits to the legitimate zone of sovereign discretion, when applied, implies that the matter falls within the legitimate zone of sovereign discretion, then how the international community ***ought*** to address/resolve the matter is not governed by sovereignty. Rather, under such circumstances, how the international community ought to act to resolve the matter is an open question, demanding a response that is not fettered by state sovereignty but is instead informed by independent moral, political and practical considerations.

This does not exhaust the possibilities. For example, there might exist a challenge that lies outside of the limits to the legitimate zone of sovereign discretion inherent to the notion of sovereignty itself, but under which there is an instrumental argument for a ***conditional*** extension of sovereign discretion. The reason for such an extension would depend on it being the case that extending sovereign discretion is—as a matter of contingent fact—the best available means to responding to the challenge under consideration. If circumstances change and/or a better alternative arises, then morality would require reform.

In the end, these concluding remarks are a conceptual bridge connecting theory to practice. A prelude to the next steps in this inquiry, and with the integrated framework now in hand and an undeveloped understanding of how the framework will be used, our attention can now turn to a more detailed explication of the framework and its application.

Part II
Making the Transition from Theory to Practice

Chapter 6
Understanding the Application of the Framework

> The good news from my point of view, is that in principle the purposes of states, and the extent of their sovereignty could be reformulated [. . .].[1]

One might be tempted to turn immediately to an application of the framework developed in Chap. 5 to assess whether the challenge of climate change falls (or ought to fall within) the legitimate zone of sovereign discretion. Doing so without further discussion of the framework or its application could lead to mistakes and/or unnecessary confusion over our understanding of the moral nature and limits of state sovereignty.

As such, this brief chapter addresses both the elements of the framework and the method of its application. It is important to note, however, that the elements of the framework are not addressed separate from the methodology, rather our understanding of the elements of the framework are developed in conjunction with our explication of the methodology. This has the virtue of avoiding any unnecessary and/or confusing conceptual analysis by focusing specifically on those aspects of the framework and its application relevant to understanding the legitimate zone of sovereign discretion.

6.1 A Threshold Consideration: Understanding the Tension Between Sovereignty and the New and Emerging Challenges Facing the International Community

Since the significance of this project depends on the existence of a moral tension, at least a *prima facie* one, between sovereignty and the new and emerging challenges facing the international community, a brief discussion of this tension is warranted.

[1]Shue (2004), p. 11.

J. J. Kassner, *Climate Change and Sovereignty*, AMINTAPHIL: The Philosophical Foundations of Law and Justice 10, https://doi.org/10.1007/978-3-030-73578-4_6

The tension is thought to arise because of a conflict between competing moral demands. On the one hand, the existing normative framework of international relations is based on the proposition that sovereignty and sovereign discretion ought to be respected. On the other hand, the new and emerging challenges facing the international community generate non-discretionary moral imperatives that are taken to conflict with sovereignty and sovereign discretion.

It is not enough to point out that there ***exists*** a moral conflict between sovereignty and the new and emerging challenges facing the international community, we also need to understand what is in conflict and why. Doing so requires that we revisit what it means to respect sovereignty. Recall, sovereignty implies sovereign discretion, which means that states are morally permitted to decide whether and how to act on those matters that fall within their sovereign discretion. In addition, respect for sovereignty requires that states refrain from interfering in those matters that fall within a state's sovereign discretion. Together, this implies that states have a wide range of moral discretion to act as they see fit and ought only to be interfered with under the most extreme circumstances.

The tension arises when this understanding of sovereign discretion is juxtaposed with the ***transnational***, ***ethical*** and ***non-discretionary*** nature of the imperatives generated by the new and emerging challenges facing the international community, like climate change. As such, understanding the tension requires that we understand the significance of these aspects of the moral imperatives that conflict with sovereign discretion.

6.1.1 The Transnational Nature of the Challenges Facing the International Community

As to the transnational nature of the challenges and the moral imperatives they generate, it is important to understand that "transnational" here means that a choice, action, or (conceptual or normative) implication extends beyond the political boundaries of a state. This simple understanding is not, however, sufficient because the transnational nature of each of the challenges can be understood from at least three different perspectives. The causes, the effects of such causes, and the actions required to respond to the moral imperatives generated are transnational. The fact that a challenge is so pervasively transnational gives rise to the tension at issue.

But why does the moral tension require that the challenges and the moral imperatives they generate are transnational? Sovereign discretion is based on the proposition that there are actions that a state can engage in that are self-regarding, actions that do not impact other morally relevant agents. If a choice of action has consequences for others, then it becomes an open question whether the protections of sovereignty apply. As such, a challenge that is transnational is one that brings into question the applicability of sovereignty.

6.1.2 *The Ethical Nature of the Challenges Facing the International Community*

Understanding the tension at issue also depends on it being the case that the imperatives generated by a challenge are ***ethical***. This is due to the fact that the tension itself is ethical. We would not be facing a moral conflict if the imperatives generated by a challenge were merely practical or prudential. If that were the case, then the moral answer would be clear, namely we should respect state sovereignty and sovereign discretion.

But what is it that makes a challenge and the imperatives it generates ethical? First, consistent with the rule of thumb in normative and applied ethics that "ought implies can," it must be the case that we can act in furtherance of the imperatives generated by the challenge being considered. If the demands implied or solutions required are beyond human agency, then it is unreasonable to claim that something ought to be done, and thus unreasonable to claim that the matter is an ethical one.

This may have as much to do with how the challenge is framed as it does with the challenge itself. For example, assuming that there are universal human rights to physical security and subsistence, consider whether this includes protection from natural disasters. If we frame this as a question about preventing natural disasters, then we are likely to conclude that this is not an ethical matter since there is very little that we can do about the occurrence of natural disasters.[2] If we frame the matter as one of responsibility for preparing for, or responding to, a natural disaster, then it is equally clear that this ***is*** a matter of ethical concern. Whether we prepare for, or how we respond to, a natural disaster is within our control.

Second, we must be able to understand the challenge from a distinctly ethical, and not merely practical or prudential, perspective. Returning to our example, one might think about the response to a natural disaster in simple logistical terms. In which case, the ethical component is missing. We could, however, also think about what it means to respect the human rights of those impacted by the disaster or how to distribute relief in an impartial and fair manner, both ethical perspectives that inform our efforts.

Third, our reason for action must be based, at least in part, on ethical considerations. Returning to the example of our response to a natural disaster, this is about the sorts of reasons that motivate our actions. If the only reasons considered are practical or prudential, such as strategic considerations, then the prescription for action is not ethical. On the other hand, if after considering all of the relevant moral reasons for action, we decide, in our best moral judgment, that we ought to act, at least in part, because of the moral imperatives generated by the human rights of those impacted by the disaster, then the prescription is sufficiently ethical.

[2]It should be noted that, due to the extent to which humans are impacting the environment, even natural disasters—at least their frequency and intensity—may soon be understood to be matters of ethical concern.

6.1.3 The Non-Discretionary Nature of the Challenges Facing the International Community

It is logically possible that a transnational ethical challenge may not undermine a state's claim to sovereignty, especially if the moral imperatives generated by the challenge are themselves ones that allow for discretion. As such, for the tension to arise the moral imperatives generated must be non-discretionary. To clarify, non-discretionary reasons are those that an agent ***must***, on pain of irrationality or moral failure, consider.

As the tension is taken to reside in the practical deliberations of states, the obvious question is whether and how the new and emerging challenges facing the international community generate such non-discretionary reasons. Since each challenge generates distinct types of moral reasons for action, specific answers to this inquiry can only be found in our analysis of each challenge. Nonetheless, for the tension to arise, the moral imperatives a challenge generates must make such non-discretionary demands on the practical deliberations of states.

The moral tension between sovereignty and the new and emerging challenges facing the international community should be clearer now. On the one hand, as a matter of convention, states have been accorded wide moral discretion. The transnational and ethical nature of the new and emerging challenges facing the international community, generate non-discretionary moral imperatives that, to be fulfilled, often require transnational (if not global) coordinated action. Put simply, we have a conflict between the moral discretion afforded sovereign states and the demand that the transnational non-discretionary moral imperatives generated by the new and emerging challenges facing the international community be fulfilled.

6.2 An Objection

Before moving on, one might object that the tension only arises if it is ***true*** that there is an ethical conflict. As such, whether a conflict exists depends on one having accurate moral knowledge of the imperatives generated by the challenges facing the international community. One might be skeptical of the possibility of such conflicts because there is sincere and reasonable disagreement over the ethical nature of the challenges under consideration, and we lack an objective standard capable of determining which of the competing ethical views is ***the*** moral truth. As such, we are either unjustified in claiming that such an ethical conflict exists, or we are engaging in an act of hubris by claiming to know that one does exist.

The objection is unjustified, at least as it relates to the argument presented here. First because the argument does not rely on a claim to such special epistemic access. In fact, the argument presented here embraces the underlying contention of the objection, that many of the disagreements surrounding the challenges facing the international community are based on competing moral claims that are sincere and reasonable. In which case, there is no available basis for determining which answer is correct. In addition, by "ethical" I do not mean that there is a single moral principle that serves as the supreme principle of morality; rather, I am adopting an approach grounded in moral pluralism and epistemic humility.

Accepting the existence of reasonable moral disagreements and moral pluralism may give rise to a different concern. The objection just considered is based on the proposition that access to moral truth is necessary for the tension to arise. Consequently, some may find the commitment to moral pluralism amidst such moral disagreement to be troubling because we are left unable to resolve ethical disagreements. This concern is based on a narrow understanding of moral pluralism. There may be forms of pluralism that are more unified than others, more capable of resolving such disagreements than others.

Nonetheless, though I am sympathetic to the concern, I remain unconvinced. For one thing, if the fact of such reasonable disagreement demonstrates a failing of ethical pluralism, it is equally problematic for other ethical theories in the face of similar real-world disagreements. Imagine that a moral disagreement exists over whether to do (X), and that both sides of the disagreement share the same understanding of the facts of the matter, differing only in their moral commitments. On one side are utilitarians with libertarians on the other. Lastly, assume that we ***must*** decide whether to do (X).

The fact that both sides have moral principles that they each believe should be controlling, only clarifies the disagreement. The fact that each side has an overarching principle to which to appeal to settle the dispute, does not settle the disagreement. The challenge is to convince others with equally sincerely and reasonably held moral beliefs that they are, in fact, wrong. Thus, as a matter of practical effect, there is no difference between ethical pluralism or any other ethical theory, at least as it relates to resolving such disputes in the real world.

Further, ethical pluralism is consistent with how we actually experience the world. Whether we are addressing interpersonal choices or policy options, we often disagree. We should adopt a perspective that accommodates this fact of reasonable disagreement. Further, pluralism presents us with a more fruitful and honest approach to the ethical disagreements we face. We experience moral conflict not as a question of what the supreme moral principle tells us to do, but as a conflict of distinct and often incomparable moral considerations. Finally, pluralism and the epistemic humility that goes along with it are consistent with the non-ideal nature of our existence. We are not omnipotent moral sages.

One should not confuse my commitment to pluralism with the idea that my understanding of "ethical" is devoid of content. For one thing, pluralism does not commit one to accepting any and all ethical points of view as being equally valid. It does not commit one to relativism. Rather, pluralism of the sort being advocated here is committed to the proposition that there are a wide range of reasonable moral principles that can conflict with one another. There are also, however, a wide range of unreasonable moral principles and points of view that are not deserving of our respect or consideration.

Pluralism, as understood here, presumes that there is an intersubjective, if not objective, reality in which there are morally better and worse ways to proceed. However, even assuming this, our ability to identify, with certainty, a single correct answer to the ethical questions facing us is a practical impossibility. When faced with questions of ethics, our epistemic and cognitive limitations leave space for reasonable disagreement. This is not to say that such disagreements will not prove challenging, yet to set certainty as our standard would be to chase a chimera instead of doing the difficult political work of constructing political and social institutions that, amidst such disagreement, produce collective choices deserving of our respect.

6.3 The Elements and Their Application: The Legitimate Zone of Sovereign Discretion

We can now return to the primary task of this chapter, providing a more detailed understanding of the elements of the framework and the method of their application. The goal here is not to restate the understanding presented in Chaps. 3 and 4, but to explicate those aspects of the limits to the legitimate zone of sovereign discretion necessary to understanding how the limits will be employed in assessing whether a challenge falls (or ought to fall) within a state's legitimate zone of sovereign discretion.

6.3.1 Zone of Sovereign Discretion Inherent to Sovereignty Itself

Recall, our understanding of the limits to the legitimate zone of sovereign discretion inherent to sovereignty itself is based on the proposition that sovereign discretion is limited by the rights of and duties owed to other members of the international moral community. In short, state sovereignty does not include moral discretion over whether to respect the rights of or duties owed to other members of the international moral community.

This understanding of the limits to the legitimate zone of sovereign discretion, depends on who or what has standing as members of the international moral community. Consequently, we must identify who the members of the international moral community are. Simply put, states are members. This may be obvious, but the significance of this proposition should not be underestimated as it plays a significant role in the discussions that follow. Assessing whether others should also be included will be discussed in greater detail in the chapters that follow.

Once we have identified who the members of the international moral community are, whether sovereign discretion is limited by the rights held by, or duties owed to, those members depends on the nature of the rights and duties in question. We must explore the moral relationships that exist between states as rights-holders and duty-bearers and the other members of the international moral community. Our understanding of these relationships determines the limits to sovereign discretion inherent to the notion of sovereignty. For example, if State A has a moral duty to respect Member B's right to X, then State A lacks the discretion to decide whether to respect Member B's right to X.

Though this is an accurate understanding of the conceptual nature of the argument and of the implications rights and other moral burdens can have on the moral discretion of a state burdened by such non-discretionary moral imperatives, it is not enough. The gap between this abstract conceptual understanding and its tangible implications for our understanding of the limits to sovereign discretion is an epistemic matter. Since the details of the moral relationships between the various members of the international moral community define the specific boundaries that make up the legitimate zone of sovereign discretion, we need a more complete understanding of those relationships.

Understanding such relationships is based on Henry Shue's account of rights and what having a right entails both for the right-holder and the duty-bearer.[3] As to the former, what it means for a right-holder to have a right, is to have an entitlement to enjoy the substance of the right.[4] That entitlement is constituted by more than just a right to be free from interference with the substance of the right.[5] One is entitled to have the substance of the right guaranteed.[6] But, what does it mean to have the substance of the right guaranteed? Shue argues that duty-bearers must provide ***social*** guarantees against ***standard threats*** to enjoyment of the substance of the right.[7]

These guarantees must satisfy three correlative burdens implicit in the proposition that to respect a right-holder's entitlement is to protect their entitlement to the enjoyment of the substance of the right. First, duty-bearers must not deprive a right-holder of the enjoyment of the substance of the right.[8] Second, the duty-bearer must protect the right-holder's ability to enjoy the substance of the right from unjustified deprivation by others.[9] Finally, duty-bearers must ensure that the substance of the right can be enjoyed by rights-holders.[10] They are socially guaranteed when social and political institutions are structured with the purpose of, and are in fact, effective at satisfying these three conditions.

As noted, an additional element of Shue's account is that the protection that a right-holder is owed is protection against "standard threats."[11] I have discussed

[3]Shue (1996).

[4]*Ibid.*

[5]*Ibid.*

[6]*Ibid.*

[7]*Ibid.*

[8]*Ibid.*

[9]*Ibid.*

[10]*Ibid.*

[11]*Ibid.*

Shue's notion of a standard threat elsewhere, reconstructing it to address some ambiguities in Shue's formulation.[12] Relying on that reconstruction,

> a standard threat to the substance of a right is one against which protection can be provided, it must be known or knowable, the relationship between the threat and the substance of the right must be direct and either intended to prevent the enjoyment of the substance of the right or with a high probability of preventing such enjoyment, and the threat must not be morally justified.[13]

With this understanding, our attention can turn to a brief explanation of the methodology to be employed in assessing the limits to sovereign discretion inherent to the notion of sovereignty itself. As noted, since this understanding of the limits to sovereign discretion depends on the rights held by and duties owed to other members of the international moral community, we must begin by identifying those others who are impacted by the challenge under consideration. Next, we must determine the moral status of those others, since not all are going to be members of the international moral community. For example, they may not be morally relevant at all, or it may be the case that their interests, though worthy of moral consideration, are not capable of generating non-discretionary moral burdens.

If they are members of the international moral community to whom duties are owed, whether and to what extent this moral fact has implications for the legitimate zone of discretion held by sovereign states will depend on two additional facts. First, we must understand what duties such others are owed; and, second, we must understand what this implies for the moral burdens borne by states. It is through understanding this relationship that we understand the limits to sovereign discretion inherent to the notion of sovereignty itself.

6.3.2 Zone of Sovereign Discretion Implied by Instrumental Value of State Sovereignty

As to the limits to sovereign discretion implied by the instrumental justification, whereas the limitations inherent to notion of sovereignty itself are conceptually inward-looking and not dependent on substantive judgments about the moral value of sovereignty, understanding the limits to the legitimate zone of sovereign discretion implied by the instrumental justification for sovereignty requires that we make such moral judgments. Simply put, since sovereignty is justified by its instrumental value, understanding the limits of sovereign discretion implied by the instrumental justification requires an assessment of the instrumental value of including a challenge within the zone of sovereign discretion.

As such, we must first understand what it means for it to be instrumentally valuable to include a challenge within the legitimate zone of sovereign discretion.

[12] Kassner (2012), p. 58.

[13] *Ibid.*

We must also consider the fact that the justification for sovereignty is its ability to serve other valuable goals within the international arena; consequently, if there is a conflict between the instrumental value of including new and emerging challenges within a state's legitimate zone of sovereign discretion and the values state sovereignty currently serves, the conflict needs to be recognized and addressed.

To understand whether it would be instrumentally valuable to include a challenge within the legitimate zone of sovereign discretion, one must understand the nature of the instrumental justification for sovereignty. As argued in Chap. 4, under the instrumental justification, sovereignty is justified by the extent to which it enhances the international community's ability to achieve valuable moral ends.

The nature of that instrumental relationship can be direct, as in the case in the relationship between sovereignty and political self-determination; or, indirect, as is the case in the relationship between sovereignty and international peace and security. Thus, the first step to assessing whether a challenge falls (or ought to fall) within the legitimate zone of sovereign discretion requires that we assess the extent to which including a challenge within a state's sovereign discretion contributes to our ability to respond to the challenge under consideration.

It is important to reiterate that the outcome of this assessment may conflict with our understanding of the legitimate zone of sovereign discretion inherent to the notion of sovereignty. Consequently, we may be faced with the additional challenge of resolving a conflict between the extant rights of, or duties owed to, some members of the international moral community and the valuable ends that might be served by intentionally extending the legitimate zone of sovereign discretion to include the challenge under consideration. How such conflicts can or should be resolved cannot be addressed in the abstract. What can be said in such circumstances, is that the evaluation should not be constrained by sovereignty. Rather, it should include an independent assessment of the relevant moral considerations in favor of and against expanding sovereign discretion.

One might wonder if the inverse is possible, a circumstance in which a challenge lies within the limits of sovereign discretion inherent to notion of sovereignty yet also determine that achieving the moral imperatives generated by the challenge may not be served by including the challenge within the legitimate zone of sovereign discretion. This sort of conflict does not present us with a similar problem. If a challenge falls within the limits of sovereign discretion inherent to the notion of sovereignty itself, then that means that the challenge and the imperatives it generates are self-regarding. As such, they would not give rise to the moral tension that makes this inquiry relevant.

It is important to remember that our understanding of the instrumental value of sovereign discretion and any limitations we might place on it, must avoid thinking of that value as being exhausted by our assessment of the instrumental relationship between sovereignty and the particular challenge at issue. Rather, we must also account for the fact that there are other morally valuable ends served by sovereignty, and the limitations we place on sovereign discretion might affect the ability of, and extent to which, sovereignty serves those other morally valuable ends.

This raises an obvious question; how will such conflicts be addressed? Again, it is not possible to provide a single answer to this question, as the resolution—if there is one available—will depend on a complicated moral assessment of the nature and relative value of the different ends being serve (or undermined), and also an evaluation of the extent to which those ends are being served (or undermined).

For example, if placing a challenge outside the legitimate zone of sovereign discretion ***greatly enhances*** our ability to address the challenge in question but will only cause the ***possibility of a minor*** harm to international peace and security, we may find this to be an acceptable moral trade-off, and thus justify placing the challenge outside the legitimate zone of sovereign discretion. Regardless, it will be important to recognize that this is an evaluation that must be addressed on its own terms.

With that said, and though this admits of a relatively flexible and open approach to such cases, there is one rule of thumb that ought to be relied upon. Namely, the analysis of such cases should begin with a presumption that those who would argue that the challenge should be placed outside the legitimate zone of sovereign discretion bear the burden of persuasion. In other words, deliberations on the matter should be governed by a presumption that harms to the extant ends being served by sovereignty are presumptively unjustified, though this presumption can be rebutted.

One might wonder why we should employ a presumption rather than treat such matters as unencumbered open questions. For one thing, the value of sovereignty as a means of promoting political self-determination and international peace and security is relatively well understood. We have ample empirical evidence of, and moral argument regarding, its value in this regard. This does not mean that the value is overwhelming, and for those who believe that sovereignty has little to do with international peace and security, if they are correct, they should find the task of rebutting the presumption that much easier.

In addition, there are also practical and pragmatic reasons in favor of tipping the scales, if only for the sake of argument, so that we can move beyond abstract debates that might become all-consuming. As the fundamental concern of this part of the project is about connecting theory to practice, a deliberative process in which the entire subject under consideration is treated as an open question is likely to lead to long and drawn out debates over exactly how to start the evaluation.

Another objection might be levied, not against the use of a presumption, but against the specific presumption in favor of privileging the relationship between sovereignty and both international peace and security and political self-determination. One might argue that we should presume that the challenge should be placed outside of the legitimate zone of sovereign discretion unless it can be demonstrated that the harm to the morally valuable ends served by sovereignty are sufficient to justify including the matter within the legitimate zone of sovereign discretion.

This is not an unreasonable alternative, but I remain unconvinced largely because of the different histories we have available to us. As noted, the role sovereignty plays in advancing the goals of political self-determination and international peace and security has been subjected to critical scrutiny. The question of whether including a new or emerging challenge within the legitimate zone of sovereign discretion would serve the morally valuable ends related to that challenge is largely speculative. Consequently, we are more likely to get the answer right if those advancing the more speculative position bear the burden of persuasion.

6.4 Institutional Design

The final step in the transition from theory to practice is to offer some suggestions for institutional design and/or reform. Though there are numerous considerations to be taken into account, the threshold matter is whether the challenge under consideration is (or ought to be) included within the legitimate zone of sovereign discretion. If the challenge falls within the legitimate zone of sovereign discretion, then the discussion of institutional design is constrained by sovereignty. If, on the other hand, the challenge under consideration is not within the legitimate zone of sovereign discretion, then how we ought to respond is unconstrained by sovereignty.

In considering what institutional designs or reforms might be appropriate, it is important to recognize that the new and emerging challenges facing the international community are wicked problems. For those unfamiliar with this notion, a wicked problem is one that is complex and complicated. The complexity of a wicked problem refers to the fact that the problem or challenge cannot be understood from a single perspective. Rather, understanding the nature of the problem is dependent on having a perspective on the matter that appreciates that numerous perspectives relevant to understanding the problem in question.

In addition to, and implicit in, such complexity is the fact that responding to wicked problems is complicated. First, the resolution of wicked problems is complicated because of the wide variety of distinct but interrelated perspectives involved in our understanding of the problem itself. Each perspective is likely to have its own standards of success and prescriptions for action. Second, in light of this complexity, it would be unreasonable to expect that a consensus will be reached on how to address the problem.

Looking to the example of poverty as paradigmatic of a wicked problem, it is an economic problem, but also a moral, political, social and historical problem. A satisfactory understanding of poverty as a problem requires an appreciation for how these various perspectives relate to, depend on and cannot be disaggregated from one another. We must understand the history of poverty along with the economic, social and political structures that have contributed to poverty. In addition, the solution is complicated. Economists disagree on the best way to address material deprivation, ethicists disagree on what would constitute a morally permissible solution to the problem, and politicians argue about how to go about achieving a solution. In addition, each of these perspectives is inexorably linked to the others, avoiding such complexity and complication is to avoid understanding the problem itself, and likely to result in institutions that are destined to fail.

Understanding the wicked nature of the new and emerging challenges facing the international community highlights the importance of social and political institutions. When faced with wicked problems, recognizing that reasonable disagreement will arise over both our understanding of the challenges faced and the solutions settled upon, it is essential that we have social and political institutions capable of performing the various epistemic, coordinating and enforcement functions in a way that is worthy of the respect of those affected by a choice of action, especially the respect of those who disagree with the choice.

Regarding this final step, there are further clarifications that warrant attention. There is an intermediary inquiry between the determination of whether the matter falls within the legitimate zone of sovereign discretion and the suggestions for institutional design and/or reform. Namely, we must determine whether there are any existing institutions worth reforming that have been charged with responding to the challenge under consideration. If the answer to this inquiry is yes, then the suggestions for institutional design and/or reform will be limited, not just by what is practically politically possible in the broader sense, but by what is practically politically possible in a narrower sense—that there is an existing institutional framework within which reform should take place.

Having addressed these threshold issues, my suggestions for institutional design and/or reform will be guided by three commitments, all of which will be discussed in detail in Chap. 9. First, the reform and/or design suggested must satisfy a minimal standard of moral and political justifiability. What constitutes a morally and politically justifiable institution will depend heavily on whether the challenge under consideration falls within the legitimate zone of sovereign discretion. If so, then respecting state sovereignty is an essential element. If not, then it is defined by a free-standing commitment to principles of moral and political justification relevant to the matter at hand.

Second, as the objective is to offer a practical response to a pressing moral concern, and not merely to provide an abstract account of the challenge or offer morally legitimate solutions for an ideal world, my suggestions will be guided by considerations of practical effectiveness. At a minimum this requires that the institution make a difference in the practical deliberations of those whose actions are governed by the institution. Third and finally, I will follow the principle of subsidiarity, the proposition that the most local institutions capable of performing a function in a morally legitimate way should be assigned responsibility for performing it.

One might wonder how a finding that a challenge is (or ought to be) within the legitimate zone of sovereign discretion makes any difference to how the international community ought to address the challenge being considered. In other words, since the extant normative framework of international relations is built upon sovereignty, the finding that a challenge being faced by the international community falls (or ought to fall) within the legitimate zone of sovereign discretion would seem to provide some support for the status quo.

This would involve both a misunderstanding of the nature of sovereignty and sovereign discretion and a conflation of what ***is*** with what ***ought to be***. As to the former, it is not the case that since a challenge falls within the legitimate zone of sovereign discretion, that states must exercise their discretion in morally good or praiseworthy ways. In fact, the point of such discretion is that states can choose for themselves whether and how to respond, including choosing to respond in ways that are morally suspect. As to the latter, with this understanding of sovereignty and sovereign discretion in hand, finding that a challenge falls within the legitimate zone of sovereign discretion does not mean that how that discretion ***is*** being used is how it ***ought*** to be used.

So, how should we approach institutional design and reform when the challenge being considered is determined to fall within the legitimate zone of sovereign discretion? To begin, we must account for the fact that whether and how a sovereign state responds is a matter of the state's moral discretion. What this means is that what counts as politically justified includes a requirement that any institutional design, and/or reform, respect sovereignty. As such, to be effective, it will be necessary to incentivize cooperative behavior and disincentivize defection.

As to the latter, it would be wrong to object that such methods are necessarily coercive and thus in conflict with sovereignty. Take as an example, those institutions where the benefits provided by the institution in question are reserved for those who cooperate. Consequently, for those states choosing to defect, they do so voluntarily as an exercise of their discretion, presumptively concluding that the benefits of defection are greater than the costs. I recognize that in actual practice the matter is not so simple, but this should suffice to make the necessary conceptual point.

6.5 Conclusion

Rather than conclude by summarizing this chapter, I will end with a brief effort to frame the discussion found in the succeeding chapters. On its surface, what follows is an assessment of whether the challenge of climate change falls (or ought to fall) within the legitimate zone of sovereign discretion. This understanding, however, fails to adequately capture the most relevant aspects of this final step in the transition from the theory to practice.

Focusing on the fact that the objective of this project is to understand the moral nature and limits of state sovereignty, the chapters that follow contribute to our understanding in two ways. First, they provide us with an answer to the specific question of whether climate change falls (or ought to fall) within the legitimate zone of sovereign discretion. Second, if we view the succeeding chapters, not simply as an assessment of whether climate change falls (or ought to fall) within the legitimate zone of sovereign discretion, but instead as an example of how to go about assessing whether any of the new and emerging challenges facing the international community fall (or ought to fall) within the legitimate zone of sovereign discretion, we will – in the end – have an understanding of how to go about gaining a more complete understanding of the moral nature and limits of state sovereignty.

References

Kassner, Joshua J. 2012. *Rwanda and the Moral Obligation of Humanitarian Intervention*. Edinburgh, UK: Edinburgh University Press.

Shue, Henry. 1996. *Basic Rights: Subsistence, Affluence, and U.S. Foreign Policy*. Princeton, NJ: Princeton University Press.

Shue, Henry. 2004. *"Limiting Sovereignty," from, Jennifer M. Welsh, Humanitarian Intervention and International Relations*. Oxford: Oxford University Press.

Chapter 7
Global Climate Change: A Matter of Transnational Ethical Concern

> The first country to industrialize and produce greenhouse gas on a grand scale, the United Kingdom, is expected to suffer least from climate change. The world's slowest developing countries, producing the least emissions, will be among those hardest hit [. . .].[1]

In September of 2019, during the United Nations' Asia-Pacific Climate Week meetings in Bangkok, several Pacific island nations met to discuss a common threat. The threat they were concerned about was (and remains) rising sea levels caused by climate change. The islands are slowly, but with certainty, disappearing beneath the rising tides. Consequently, due to the territorial nature of state sovereignty, their legal and moral standing within the international community is literally being drowned beneath the sea.

Admittedly, the existential threat rising sea levels pose to these island nations is just one amongst many of the harms of climate change, an illuminating microcosm of the challenges the world is facing. And though one might want a fuller accounting of the ways in which climate change is causing harm around the globe, those harms will not be catalogued here. For one thing, the list itself would quickly become overwhelming. More importantly, that effort would only have limited value to the project at hand.

Nonetheless, the existential threat that rising sea levels posed to the island nations that met in Bangkok succinctly captures those aspects of climate change that make it so well-suited to serve as a test case in our efforts to understand the moral nature and limits of state sovereignty. To be specific, understanding the moral nature and limits of state sovereignty in light of the new and emerging challenges facing the international community is inexorably linked to a number of defining characteristics of the challenge of climate change. Namely, climate change has been brought about by the actions and omissions of individuals, corporations and political communities around the world, resulting in harms suffered in every corner of the globe, often suffered by people and political communities who contributed little to the underlying cause(s).

[1]Wells (2019), p. 132.

J. J. Kassner, *Climate Change and Sovereignty*, AMINTAPHIL: The Philosophical Foundations of Law and Justice 10, https://doi.org/10.1007/978-3-030-73578-4_7

Climate change and the harms it has caused are thus a direct result of the ways in which our world has become deeply interconnected and interdependent.

In addition, even assuming the best of intentions, individual states acting on their own cannot adequately—morally or practically—respond to the challenges posed by climate change. Rather, to effectively respond to the challenge of climate change in a morally and politically justified way, the international community will have to act cooperatively in highly coordinated and deliberate ways. In the end, what makes climate change an appropriate place to explore the implications the theory developed thus far has for the moral nature and limits of state sovereignty is that there may be no challenge more antithetical to the discretion afforded sovereign states.

One might wonder how much can possibly be discovered about the moral nature and limits of state sovereignty by assessing a single challenge like climate change. It is useful to remember that our understanding of the moral nature and limits of sovereignty cannot be determined categorically. Rather, as has been discussed, due to the instrumental nature of the moral justification for sovereignty, our understanding must be constructed through a piecemeal assessment, determining the moral limits of state sovereignty through independent inquiries assessing whether each of any number of challenges falls (or ought to fall) within the legitimate zone of sovereign discretion.

Nonetheless, and despite the fact that assessing whether climate change falls within the legitimate zone of sovereign discretion cannot exhaustively determine the moral limits of state sovereignty, a rigorous application of the framework developed in the preceding chapters to determine whether climate change falls (or ought to fall) within the legitimate zone of sovereign discretion serves as both a limiting and an instructional case. Limiting because it is one of the most obvious examples of a challenge facing the international community in tension with state sovereignty. Consequently, if we conclude that climate change falls within the legitimate zone of sovereign discretion, then it is likely that all other new and emerging challenges facing the international community would fall within the legitimate zone of sovereign discretion as well.

It is instructional because in applying the framework to a challenge like climate change, we are presented with, perhaps, the simplest case to be addressed. This simplicity allows us to see the tension that exists between the challenge of climate change and the moral discretion afforded sovereign states clearly, providing a model that can be used to ground our assessment of whether other challenges facing the international community fall (or ought to fall) within the legitimate zone of sovereign discretion, in which the tension may not be readily apparent.

As was argued in Chap. 6, for this inquiry regarding climate change to be relevant, it must be the case that a conflict exists between climate change and the moral imperatives it generates and sovereign discretion. Whether such a conflict exists depends on whether the challenge under consideration and the imperatives it generates are transnational, ethical and non-discretionary. As such, the remainder of this chapter will establish that climate change and the moral imperatives it generates are transnational, ethical and non-discretionary and thus in conflict with the moral discretion afforded sovereign states.

7.1 The Transnational Nature of Climate Change

Recall, to generate the tension, "transnational" means a challenge that is pervasively transnational. Though what is required cannot be defined with exact specificity, at the very least, what is required is that the challenge cannot be adequately understood without an appreciation of the fact that it is transnational. In other words, the ways in which the challenge is transnational must be relevant to our moral and practical understanding of the challenge itself. In the case of climate change, the matter is overdetermined. The transnational nature of the challenge of climate change permeates nearly every aspect of our understanding of climate change. The causes of, effects of and solutions to climate change are all transnational.

As to the causes of climate change, no single state or its population is individually responsible for causing climate change, rather climate change is caused by the cumulative actions and omissions of individuals, corporations, and political communities from around the world. In addition, climate change is caused in an indirect way, enabled by the policies and regulations of political and legal institutions at every level. From local land use policies to national emissions standards to regional and international treaties intended to regularize international environmental law, such policies and regulations are determinative of the normative framework governing the activities of nearly every agent—individual or collective—around the globe. It is those policies that frame, incentivize and enable the actions of nearly every agent—individual and collective—around the globe that contribute to climate change.

Unsurprisingly, the effects of climate change are transnational as well. The actions and omissions of a political community and its population in one place have dramatic, and often catastrophic, consequences for people and political communities around the world. Whether one is referring to rising sea levels, the loss of biodiversity, or the threat to international peace and security posed by a changing climate, the effects of climate change are, and will for a long time be, felt in every corner of the globe.

The fact that such harms are ubiquitous, however, does not adequately capture the transnational nature of the harms of climate change. It is not just that across the globe each state and its population suffer the effects of their own actions. Rather, the effects of climate change are transnational negative externalities that don't respect political borders. I will return to this idea below, but for the moment an example may help to clarify the way that the harms of climate change are transnational.

Let us consider, for the moment, the fact that when the United States decides on emissions standards for its coal-fired power plants or chooses to direct funding to road construction over commuter railways, these political choices cause climate related ripples that spread across the globe. These impacts are unaffected by the international political boundaries that separate one sovereign political community from another. In addition, the consequences of such choices endure over decades, if not longer. In short, the harms caused by the choices of the U.S. government are not internalized.

Lastly, climate change is transnational in the sense that an adequate response to climate change requires global action. This is implied, in part, by the fact that the cause of climate change is global. This is, however, also supported by the fact that a legitimate institutional response to climate change requires a highly coordinated transnational effort. Though I will return to discuss what constitutes an "legitimate" response in greater detail in Chap. 10, what it means for a response to be "legitimate" requires a brief discussion to insure against any unnecessary misunderstandings.

One might take "legitimate" to be primarily concerned with practical matters, that the response is legitimate if it effectively addresses the challenge. Though an important aspect of a legitimate response, that would fail to account for the fact that what constitutes a legitimate response to any challenge facing the international community must address more than just the practical aspects of the challenge. Climate change is as much a political and moral problem as it is a practical and scientific one. Thus, what counts as a legitimate response is one that is not only practically effective, but also politically and morally justified.

One final point warrants a brief mention before discussing climate change as a matter of transnational ***ethical*** concern. In assessing the extent to which the challenge of climate change is transnational, one should not conclude that this has logical implications for how we ought to assess moral responsibility for climate change. The fact that climate change is transnational neither absolves individual states of their unique moral responsibility for climate change or its effects, nor does it single out which actors ought to be held responsible and to what degree. In identifying the challenge of climate change as transnational, all that can be concluded is that the causes of, effects of and responses to climate change are transnational.

7.2 Climate Change As a Matter of Ethical Concern

Having established the transnational nature of climate change, understanding the ethical tension that exists between climate change and state sovereignty requires that we clarify the ethical nature of the challenge of climate change. As an ethical matter, the moral tension upon which the relevance of this inquiry depends arises when one considers the impacts the moral imperatives generated by climate change and the moral discretion of sovereign states are taken have on the practical deliberations of states. And, for that tension to constitute a conflict, it must be the case that the moral imperatives generated by climate change include non-discretionary moral reasons for action that conflict with moral discretion afforded sovereign states.

7.2.1 Climate Change As an Ethical Matter

First, before discussing the moral nature of climate change and the moral imperatives it generates, we must address a more basic question, one that determines whether this inquiry is meaningful or moot. We must determine whether climate change is a matter of ethical concern at all. For many, the answer may seem obvious, but a failure to address this question leaves the entire edifice that is built upon its answer open to criticism. In addition, the importance of this question is only made more apparent when we consider the objection of the climate skeptic. It is worth discussing the underlying concern in more abstract terms before turning to that discussion.

Simply put, any inquiry that seeks to determine what ought, as a moral matter, ***be done*** requires that an action or state of affairs to be a matter of ethical concern. At the very least, and as was argued previously, the object of our inquiry must be amenable to human agency. For example, a state of affairs is a matter of ethical concern if it was brought about or could be altered by human action. It would make no sense to claim that someone ***ought*** to do something in a case where they are powerless to do anything about the matter at hand.

Thus, for the challenge of climate change to be a matter of ethical concern, it must be the case that there is something that we, as human agents, can do about climate change. If not, then it is not a matter of ethical concern, at least not from a prescriptive point of view though we may be able to judge it as being good or bad. As to this last point, it is possible for us to recognize that a harmful event has occurred that is beyond our control while also recognizing that those who are suffering from the event are being harmed in a way that merits a moral description of their suffering. In other words, the fact that we can describe the state of affairs in a morally meaningful way does not mean that moral prescriptions are available.

As to the need to establish that climate change can be addressed through human action, it is important to begin with a fuller discussion of the distinction just introduced. On the one hand, there is the nature of the cause of climate change. Is climate change anthropogenic? This question must, however, be distinguished from the fact that, regardless of our understanding or position on the anthropogenic nature of climate change, how we ought—as a matter of morality—respond to the effects of climate change is a conceptually independent matter. Put more succinctly, even if climate change is not anthropogenic, that does not absolve us of moral responsibilities related to the effects of climate change, whatever its causes. Even if humans can't do anything about the cause(s) of climate change, we can respond to the suffering caused by climate change.

The ethical significance of this distinction can be seen in the challenge of the climate skeptic. The skeptic who asserts that climate change is not anthropogenic, and thus beyond the reach of human control, may contend that there is little we can do about the ***causes*** of climate change, thus rendering the cause(s) of climate change morally irrelevant; however, and regardless of the truth of the skeptic's claim, the same claim cannot be made in response to the proposition that whether and how we ***respond*** to the impacts of climate change is an ethical matter.

The control we have over the causes of and responses to the harms of climate change are related in significant rhetorical and practical ways, but the point is a conceptual one. The fact that the cause of a harm is beyond our control does not imply that the response to that harm, once caused, is beyond our control. Though knowing what caused the harm may help us to understand how to respond in the most effective way, the fact that we—in the skeptic's view—can do little about the cause does not imply that we are powerless to offer aid, comfort or support in response.

With that distinction in mind, is the challenge of climate change responsive to human agency? There is an overwhelming scientific consensus that climate change is anthropogenic,[2] providing us with a *prima facie* reason to presume that humans are causing climate change and are thus able to respond to the causes of climate change. There are, however, at least two potential objections to be considered. First, one might argue that, much as you cannot step in the same river twice, no matter what we do, recovering the environment we once had is not possible. Second, there are many who, despite the overwhelming scientific consensus, remain skeptical of the proposition that human activity is causing climate change.

As to the first concern, stopping climate change and recreating a historical state of climatic and environmental affairs identical to that which existed prior to the era of anthropogenic climate change is a practical impossibility. That we cannot recreate a specific historic environmental situation does not, however, mean that we are incapable of doing anything about the human activities that are causing and worsening climate change. We could focus on halting and/or reversing the anthropogenic drivers of climate change, rather than trying to recreate an environmental situation that no longer exists. Though different in morally relevant ways, this would be a meaningful moral endeavor. In the end, accepting that climate change is anthropogenic means that some action to address the causes of climate change is within our control.

7.2.2 Responding to the Climate Skeptic

The arguments just presented are meaningless for the climate skeptic. There are several arguments, however, that weigh decisively against the climate skeptic's position. The first begins by recognizing that there is an important distinction between our empirical/scientific understanding of the causes of climate change and the question of whether climate change is a matter of ethical concern, especially in terms of the latter's relationship to public policy. In short, we need to separate the empirical inquiry with which scientists are concerned from the ethical inquiry with which we are concerned.

The scientific inquiry focuses on identifying the empirical facts of the matter and offering an understanding of those facts and related phenomena. On the other hand, when it comes to our ethical/policy analysis, our primary concern is with making the best possible (policy) choice we can, based upon the information available to us. We

[2]Intergovernmental Panel on Climate Change Assessment Reports.

need to make choices in the face of uncertainty and cannot let the desire for absolute certainty be an obstacle to making good (policy) choices.

Noting this distinction, let us assume for the sake of argument that there is a controversy over the scientific claim that climate change is anthropogenic. On one side of this "debate" we have an overwhelming consensus that climate change is anthropogenic. On the other, we have a smattering of scientists—many who are not experts in climate or environmental geophysics—who doubt the anthropogenic nature of climate change. Consequently, on the factual matter alone, the skeptic has good reason to be humble in their assessment of the causal bases for climate change and should, at the very least, be open to the possibility that they are wrong.

Additionally, there is an important difference between the significance of uncertainty in the quest for scientific knowledge versus the significance of uncertainty when public policy choices are at issue. As to the former, the quest for scientific knowledge is taken to be valuable in-and-of-itself. This is one of the main intellectual motivations of basic research. The objective is to move to ever more accurate understandings of ourselves and our world. As such, uncertainty is a clarion call for more inquiry and investigation.

When, however, we are facing a public policy choice that depends on our empirical understanding of phenomena, our objective is not, and ***should not*** be, primarily concerned with seeking out a complete and unassailable understanding of the relevant phenomena. This is not to say that we wouldn't want to have as accurate an understanding of the phenomena as is practically possible, but our primary concern is and ought to be with the policy choice we face. Specifically, considering the evidence available to us, what policy choice(s) ought to be pursued? Our scientific understanding is instrumental to making the best policy choice we can, but a scientific understanding is not, and should not be, the ultimate goal being sought by policymakers.

One may think that this leaves us without determinative reasons for choosing one perspective over another. This is mistaken, there are good reasons for presuming that climate change is anthropogenic, even in the face of uncertainty. For one thing, our political leaders have special obligations to protect their populations from harm. The risks of harm, should we choose not to act because of the climate skeptics' demand for epistemic certainty, are simply too great. Whereas, the opportunity costs that would come from combatting climate change and pursuing more environmentally friendly and sustainable policies would be minimal and would likely produce compensating benefits apart from the impact our efforts may have on climate change. For example, we would likely end up with more efficient cars, cleaner air and less dependence on dictators and tyrants for our fuel.

In addition, whether, as a policy matter, we ought to ***treat*** climate change as anthropogenic depends on the probabilities associated with the possibility that climate change is anthropogenic. Taking the scientific consensus into account, there is a much greater likelihood that climate change is anthropogenic than not. Consequently, the moral burden borne by our leaders is only clearer. To reject this consensus is to act against known probabilities, adding to the failures of policymakers who choose—against all evidence to the contrary—to remain skeptical of, and make policy choices based on their skepticism over, the anthropogenic nature of climate change.

For the sake of argument, let us set aside the overwhelming scientific consensus and the morally irresponsible failure to recognize the potentially catastrophic risks that accompany inaction based on the climate skeptic's beliefs. And instead, let us assume that climate change is not anthropogenic, and that it is beyond the capabilities of humankind to respond to the causes of climate change. Though this may render such causes beyond ethical concern, ***how we respond*** to the impacts of our changing climate is a separate matter. Our response to the harms of climate change is within our control.

An analogy may help to clarify this point. Imagine a natural disaster like a tsunami, hurricane or earthquake, the natural disaster itself is not subject to human control.[3] Since it is beyond the reach of human agency, its cause is ***not*** a matter of ethical concern. What we do in response, both in terms of post-disaster relief and policies pursued to protect against the possible harms of future natural disasters, however, ***are*** within our control and are thus matters of ethical concern. Thus, even operating under the assumption that climate change is sufficiently similar in relevant ways to other natural events, though we can't influence such events, we have control over how we respond to the harms they cause. Consequently, whether and how we respond ***is*** a matter of ethical concern regardless of the anthropogenic nature of climate change.

Though answering the skeptic, this would change the ethical nature of the challenge of climate change. If we are focused on the causes of climate change ***and*** the possible responses to the harms of climate change, since this implicitly assumes the anthropogenic nature of climate change, questions of moral responsibility for past actions and moral responsibility for future efforts at mitigation matter. If, however, our understanding of climate change as an ethical matter is limited to whether and how we respond to the harms of climate change, such moral considerations are irrelevant, and our focus would then shift to assessing the extent to which we have duties to aid distant others. For many, this would mean a shift from a moral understanding grounded in claims of justice to one grounded in claims of charity or benevolence—a distinction that I have rejected elsewhere.[4] Nonetheless, we cannot avoid the fact that how we ought to respond to the suffering caused by climate change is a matter of ethical concern.

Notwithstanding this, there are two compelling reasons for accepting that climate change is anthropogenic. As noted above, as a matter of probability, it is far more likely that climate change is anthropogenic than not. Further, considering the risks involved, if climate change is anthropogenic and we do nothing about it the likely

---

[3]As least with regards to severe weather events like hurricanes, there is a growing scientific consensus that climate change will, if it hasn't already, increase the intensity and frequency of such "natural" events. https://nca2014.globalchange.gov/highlights/report-findings/extreme-weather.

[4]Kassner (2009).

outcome is catastrophic. If, on the other hand, it is not anthropogenic, then our efforts to halt or mitigate the extent to which human activities contribute to climate change will have no effect on climate change, but there are benefits that would accompany such efforts. In the end, we should—and this discussion will—operate under the assumption that climate change is anthropogenic.

7.3 Moral Considerations and Practical Deliberations

In addition to the fact that responding to the challenge must be within the scope of human agency, for a challenge to be an ethical one, we must be able to understand the challenge and the imperatives it generates from a distinctly ethical point of view. This requires that the underlying moral values and considerations that define the challenge must be ethical, and the moral imperatives to act must be the outcome of practical deliberations wherein moral considerations played a decisive role in the determination of what ought to be done. In addition, those moral imperatives must be non-discretionary, in which case an agent who is deliberating will have deliberated wrongly if they have failed to consider the moral imperatives generated.

7.3.1 Identifying the Moral Considerations Relevant to Climate Change

The fact that global climate change is an ethical challenge that, morally speaking, demands action does not tell us what form that action should take. Consequently, in articulating the ethical nature of the challenge of climate change, one must also explain what moral considerations are implicated ***and*** the nature of the moral prescriptions they generate.

The central moral claim is that climate change involves various morally problematic harms to individuals and groups. Thus, we are faced with two distinct issues to be addressed. First, we must identify harms to individuals and groups that can be attributed to climate change. And, second, we must justify the proposition that such harms are morally problematic.

Before proceeding, there is one conceptual point that requires clarification, even if only by means of definition. Namely, since we might disagree about what constitutes harm, it is important to be explicit about what is meant here. Though likely unsatisfying to some, harm here is used in a colloquial sense. An agent or entity is harmed if it is made materially worse off by an action or event. This is a non-moral notion of harm and is merely an expression of the empirical facts of the matter. Thus, it is possible to distinguish between morally permissible and morally problematic harms.

Thus, as to the first task before us, the identification of the harms at issue is an empirical matter. The claim that climate change is causing harm cannot be justified by conceptual analysis or moral argument; rather, it depends on it being empirically true that climate change is causing harm. Consequently, it will be necessary to rely on a few real-world examples to establish the truth of the underlying proposition. As such, and to return to the example that began this chapter, we can look to the immediate harms being suffered by many island nations and coastal communities around the globe. One of the most clearly documented impacts of climate change is the threat posed to such communities by dramatically higher seas, flooding coastal communities and rendering island nations permanently uninhabitable—all caused by climate change.

To avoid the mistaken impression that rising seas is the only harm worthy of our concern, another example of the harms of climate change can be found in the consensus around the proposition that climate change is a major contributor to weather disruptions around the world, including both desertification and flooding across the African continent.[5] There, climate change has disrupted hydrologic cycles, causing drought and flooding undermining political stability and exacerbating other challenges faced by already vulnerable populations.[6]

These examples provide us with the empirical bases to demonstrate that climate change is causing harm, suffered by both individuals and groups. As to the harms affecting individuals, such harms include—though not limited to—the fact that the homes and communities of those who live in coastal areas and on island nations are being rendered uninhabitable by rising sea levels.[7] Such material and personal losses will only get worse as the effects of climate change intensify. In addition, the fact that climate change has led to the desertification of formerly arable land across the continent of Africa,[8] has led to food insecurity and left individuals susceptible to starvation and disease.[9]

As to the harms to groups caused by climate change, perhaps most relevant to this discussion is the threat climate change poses to political communities. As noted, rising sea levels pose a threat to the habitability of various island nations; consequently, the very territorial basis for the existence of those political communities is under threat.[10] In a world where political authority, jurisdiction and sovereignty are all territorially defined, it would be difficult to overstate the significance of such a loss. Similarly, the disruption of weather patterns resulting in floods and

---

[5]Barnosky (2009), pp. 121–131.

[6]See Barnosky (2009) and see Wells (2019).

[7]Randy Astaiza, "Islands Threatened by Climate Change", http://www.businessinsider.com/islands-threatened-by-climate-change-2012-10.

[8]Wittig et al. (2007), p. 182.

[9]Scott Baldauf, Africans are already facing climate change", Christian Science Monitor (Nov. 6, 2006). https://www.csmonitor.com/2006/1106/p04s01-woaf.html.

[10]Roger-Mark Desouza, "Paris Climate Deal: Island Nations Face Extinction From Rising Seal Levels", Newsweek (June 10, 2017). http://www.newsweek.com/trump-paris-climate-deal-hawaii-sea-levels-climate-change-623500.

desertification in Africa and the pressure this has placed on the ability of local communities to provide for themselves has led to migration and political instability, undermining political self-determination and harming national and regional security.[11]

As noted, understanding that the effects of climate change are causing harm does not, without more, establish that such harms are morally problematic, or, even if they are, what moral prescriptions follow. Consequently, it is necessary to explain why the harms in question are morally problematic and what, if any, prescriptions follow. As to the former, it is largely uncontroversial proposition that if one's actions harm another, then we ought to presume that the former has wronged the latter unless a justification can be provided.

To further understand the nature of the harms of climate change we should begin by recognizing that such harms are not self-regarding. Due to the fact that climate change is transnational in both its causes and effects, those who suffer the ill effects of climate change cannot avoid them. As to the moral significance of the proposition that the harms of climate change are not self-regarding. It is important to note that I don't mean to imply that those whose activities have caused climate change don't suffer any of the harms of climate change. We all do. That can't be avoided. Rather, by "not self-regarding" I mean that the harms ***can't be confined*** to those who have produced them. As noted earlier, borrowing from economics, the harms are negative externalities.

One important moral implication of the fact that the harms of climate change are not self-regarding is that one cannot defend inaction under a Millian form of anti-paternalism.[12] The harms are external to the agent causing them; consequently, one cannot rely on the proposition that an agent's freedom should not be limited if the agent is only harming themselves. For the moment we can set aside the fact that states are not analogous to individuals and can reject such anti-paternalist arguments as entirely inapplicable and not merely disanalogous.

Additionally, the harms of climate change are ubiquitous and indiscriminate, harming groups and individuals who have no opportunity or ability to avoid the harms in question or object—in a meaningful way—to the actions and omissions causing climate change. This is because the normative framework of international relations is built around a commitment to the territorially defined sovereign state. One implication of this commitment it that the choices made in one state that are contributing to climate change and causing harms to those residing in another are, at least as a conventional matter, beyond the political reach of many who are harmed by such choices. Thus, the harms of climate change are forced upon individuals and groups without their consent. This amounts to using the effected individuals and groups as "mere means" for one's own purposes, violating the Kantian principle that

---

[11]Scott Baldauf, Africans are already facing climate change", Christian Science Monitor (Nov. 6, 2006). https://www.csmonitor.com/2006/1106/p04s01-woaf.html.

[12]Mill (1991).

one has a duty to not use others as a mere means, but must respect them and their humanity as ends in themselves.[13]

Though some may find it unsatisfactory, this is the extent of my discussion of the underlying Kantian principle, and I will rely on the extensive work by others for its defense.[14] Nonetheless, and regardless of whether one agrees with the Kantian understanding of the underlying claim, it is a moral proposition that fits well with our common sense morality and is one that I find convincing. The success of the argument, however, does not depend on this principle, but it is supported by it. In the end, it should be readily apparent that the harms of climate change are morally problematic. Harms that morality ***demands*** be addressed.

One might object that the moral implications of climate change are not so simple, that the activities that have caused and/or contributed to climate change have produced and continue to produce benefits alongside those harms. For example, the production of greenhouse gases was a necessary, and arguably unintended, consequence of the industrial revolution's dependence on coal and oil. One might further argue that the developing world now depends on the same fuels for its own economic improvement. In short, the harms of climate change must be understood within a broader context that also accounts for the benefits the causes of climate change have produced and will produce.

It is undeniable that many have benefited from the activities that contributed to climate change. The prosperity of the developed world would not have been possible. Even today, it would be a denial of the obvious not to recognize that the aspirations of many in the developing world are built around the use of cheap fuels to support their economic development. Such benefits do not, however, negate the harms caused by the activities producing them. Recognition of such benefits is one of many additional moral considerations that must be taken into account when identifying a morally justifiable solution to the challenge of climate change, but they do nothing to morally absolve us of the harm our activities cause, or to justify our choice to continue on with "business as usual."

Before moving on, it is important to note that, even under the most charitable interpretation of this objection, it is merely a partial one. For one thing, the choices of those capable of being held morally responsible for their actions, namely paradigm adults, are the immediate beneficiaries of their choices. Yet, the harms will not be felt by those who made those choices or reaped the benefits. Rather, the harms will fall on future generations, both existing and non-existent. Consequently, the moral imperative to do something about such harms cannot be avoided by pointing to the benefits produced by the underlying causes of climate change.

In addition, at present, the primary beneficiaries of the activities that caused climate change are those in the developed North, while the harms are being suffered most by those in the developing South. This asymmetry is morally relevant and weighs heavily against the persuasiveness of the objection. It would be akin to a

---

[13]Kant (1993).

[14]See Baron (1999). And see Kerstein (2002).

slaveowner seeking to justify the harms of his use of slave labor by appealing to the fact that it benefitted him (the slave owner) greatly.

7.3.2 A Matter of Practical Deliberations: The Non-discretionary Nature of the Moral Imperatives Generated by the Challenge of Climate Change

As discussed in Chap. 6, the significance of this project depends on the challenge under consideration generating non-discretionary moral imperatives, moral demands that cannot be ignored or set aside within the practical deliberations of the agent to whom they apply. It is such non-discretionary moral imperatives that are in conflict with the moral discretion sovereign states are presumed to have. As to the challenge of climate change, the relevant moral considerations are the duty not to harm and the duty not to treat others as mere means.

As for understanding the moral imperatives generated by the duty not to harm, I am relying on a simple yet powerful argument from Thomas Pogge. Pogge argues that we all have a duty not to harm others.[15] When we violate that duty, including by participating in institutions or practices that lead to such unjustified harm to others, we have an obligation to stop engaging in the harmful actions and to provide rectification for past harms.[16] Most importantly, the resultant moral obligations are not matters of beneficence or charity, rather they are matters of the most basic justice.[17] We ***must*** consider such moral imperatives in our practical deliberations.

Thus, implicit in the duty not to harm is a primary non-discretionary moral duty to refrain from causing or participating in activities that cause harm to others. In addition to such primary duties, remedial moral responsibilities arise when an agent fails to fulfill their primary responsibilities under the duty not to harm. Under these remedial responsibilities, one is under a moral obligation to stop engaging or participating in activities that cause morally unjustified harm to others ***and*** have a non-discretionary moral obligation to act to rectify the past harms to which one has contributed.

As to the duty not to treat others as a means, due to the fact that those suffering the effects of climate change have no opportunity to consent to, dissent from or participate in the decisions and decision-making processes that led to the harms they are suffering. They have been treated as mere means. Here too, those responsible for such wrongful action have a moral duty to stop engaging in the underlying behavior. In addition to the remedial moral responsibilities that follow from such violations, there may also be duties of justice related to the establishment of social and political institutions that provide those affected by climate change with

---

[15]Pogge (2008).

[16]*Ibid.*

[17]*Ibid.*

meaningful opportunities to participate in the decision-making processes relevant to the causes of or responses to climate change, to insure that they are treated s ends-in-themselves.

7.4 Concluding Remarks

In the end, there seems to be little reason to doubt that climate change is a matter of transnational ethical concern generating non-discretionary moral imperatives. Whether one is discussing what ought to be done to combat the causes of climate change or how the international community ought to respond to the harms caused by climate change, both can be influenced and/or addressed through human action. Since the harms of climate change are negative externalities and not purely self-regarding and are forced upon a powerless global population—violating both the duty not to harm and the duty not to use others as mere means—they are moral wrongs generating non-discretionary moral imperatives that conflict with sovereign discretion, a conflict that makes the next steps in our effort to make the transition from theory to practice relevant and meaningful.

References

Barnosky, Anthony D. 2009. *Heatstroke: Nature in an Age of Global Warming*, 121–131. Washington, DC: Island Press.

Baron, Marcia. 1999. *Kantian Ethics Without Apology*. Cornell: Cornell University Press.

Kant, Immanuel. 1993. *The Groundwork for the Metaphysics of Morals*. 3rd ed. Indianapolis: Hackett Publishing.

Kassner, Joshua J. 2009. Completing the Incomplete: A Defense of Positive Obligations to Distant Others. *Journal of Global Ethics* 5 (3): 181–193.

Kerstein, Sam. 2002. *Kant's Search for the Supreme Principle of Morality*. Cambridge: Cambridge University Press.

Mill, John Stuart. 1991. *On Liberty (1859) from On Liberty and Other Essays*. Oxford: Oxford University Press.

Pogge, Thomas. 2008. *World Poverty and Human Rights*. 2nd ed. Cambridge: Polity Press.

Wells, David Wallace. 2019. *The Uninhabitable Earth: Life After Warming*, 132. New York: Tim Duggan Books.

Wittig, R., K. König, M. Schmidt, et al. 2007. A Study of Climate Change and Anthropogenic Impacts in West Africa. *Environmental Science and Pollution Research-International* 14: 182. https://doi.org/10.1065/espr2007.02.388.

Chapter 8
Applying the Framework

In Chap. 7, I established that the challenge of climate change is transnational and that it generates non-discretionary moral demands on the practical deliberations of states and the international community. As such, there is, at the very least, a *prima facie* moral conflict between the moral discretion afforded sovereign states and the moral imperatives generated by the challenge of climate change. It bears repeating that if there was no conflict, there would be no competing demands on the practical deliberations of states giving rise to a moral tension in need of resolution, thus there would be no need to inquire further.

It is this tension that generates the questions central to the broader inquiry with which this project is concerned. Of particular importance are those questions implicating the practical deliberations of states. For example, do states have moral discretion over the challenge of climate change, thus alleviating the practical deliberations of states from the burdens imposed by the moral imperatives generated by climate change, rendering the tension irrelevant as a matter of practical deliberation? Or does the challenge of climate change lie outside of the legitimate zone of sovereign discretion, thus implying that states lack moral discretion when it comes to climate change? In which case, states and the international community ***must*** consider and properly weigh the moral imperatives generated by climate change and cannot rely on sovereign discretion to avoid this moral burden.

In this and the next chapter, these questions are answered by applying the framework developed in Chap. 6. Pursuant to that framework, this chapter determines whether climate change falls within the limits to the legitimate zone of sovereign discretion inherent to the notion of sovereignty itself. In the end, I argue that the challenge of climate change falls outside of the limits to the legitimate zone of sovereign discretion inherent to the notion of sovereignty. Chapter 9, then assesses whether there is an instrumental justification for including the challenge of climate change within the legitimate zone of sovereign discretion.

J. J. Kassner, *Climate Change and Sovereignty*, AMINTAPHIL: The Philosophical Foundations of Law and Justice 10, https://doi.org/10.1007/978-3-030-73578-4_8

8.1 Introduction

Recall that in determining whether climate change falls (or ought to fall) within the legitimate zone of sovereign discretion inherent to the notion of sovereignty itself, the assessment is based on the proposition that sovereign states are rights-holding and duty-bearing members of the international moral community. As such, choices and/or actions of states that violate the rights of, or duties owed to, other members of the international moral community are outside of the legitimate zone of sovereign discretion.

Thus, assessing whether the challenge of climate change falls within the limits to the legitimate zone of sovereign discretion inherent to state sovereignty unfolds in three steps.. First, we must identify those who are being harmed by climate change. Second, we must determine whether those harmed have standing (as either right-holders or the beneficiaries of primary duties owed) as members of the international moral community. Third, and finally, we must assess the extent to which the harms caused by climate change constitute failures to respect the duties owed to those harmed.

As to the identification of those harmed, the analysis here is limited to individuals and states. Some may find this unjustifiably narrow as there are likely many others who may have moral standing within the international moral community.[1] Though I am sympathetic to this view, the choice to rely on this narrower understanding is grounded on two considerations. First, arguing for a broader membership in the international moral community invites controversy. And, second, this narrower understanding is all that is needed to assess whether the actions (or omissions) taken by states that contribute to climate change violate the rights held by or duties owed to other members of the international moral community.

8.2 Those Harmed by Climate Change

We can quickly dispense with an implicit threshold question of whether climate change causes harm. The case is overdetermined. As was demonstrated in Chap. 7, there is little reason to doubt that individuals and states suffer significant harms caused by climate change. Thus, we can turn to the question of whether those suffering such harms have moral standing as members of the international moral community. I will begin by discussing whether states have standing as members of the international moral community because the argument in defense of the proposition that they do is largely uncontroversial.

To be specific, that sovereign states have moral standing within the international moral community is an implication of the proposition that all states are rights-holding members of the international moral community. Denying this would

[1]See Beitz (1999).

undermine the very idea that sovereignty could generate the sort of moral discretion states claim within the normative framework of international relations. In other words, the state system is built upon the proposition that states have rights that operate, not as Hohfeldian liberties, but as claims, powers and immunities, and each state has an equal claim to these rights. Consequently, the normative framework of international relations is built upon the proposition that if one state is a rights-holder within the international moral community, then all states are. To reject this proposition is to reject the state system itself.

As to the moral standing of individuals, the answer is more complicated because the moral standing of individuals is not implied by any commitments inherent to the state system. It is logically possible to imagine a circumstance in which states have moral standing as members of the international moral community, while individuals do not. Thus, a free-standing argument is needed to demonstrate that individuals have standing within the international moral community. On this point, there are several reasons that weigh decisively in support of the inclusion of individuals as members of the international moral community.

First, such standing is presumed both legally and rhetorically in the normative framework of international relations. Consequently, as a matter of social fact, individuals have standing within the international moral community. In support of this proposition, looking first at the rhetoric of international political discourse, there is a clear commitment to the standing of individuals, which can be observed when states criticize one another for failing to respect human rights and human dignity, and similarly when states tout their respect for human rights and human dignity as a point of pride.

As to the claim that, as a matter of international law, individuals are treated as though they have standing, human rights protections are recognized, if not always respected or enforced, as valid juridical norms that ***ought*** to govern the behavior of states. Whether one is focused on the activities of various domestic courts or on the numerous international tribunals tasked with upholding human rights—the European Court of Human Rights, the International Criminal Court, the Inter-American Court of Human Rights, or the African Court of Justice and Human rights, among others—the cases heard, arguments presented and decisions reached are matters of serious concern. Simply put, states seek to avoid being found responsible for human rights violations.

In addition, and though straddling the conceptual line between law and rhetoric, few would doubt the significance of the Universal Declaration of Human Rights (UDHR) and the various conventions that have been enacted to bring the aspirations of the UDHR into effect.[2] Similarly, and though a nascent and flawed doctrine, the

[2]International Convention on the Elimination of All forms of Racial Discrimination (Dec 21, 1965); International Covenant on Civil and Political Rights (Dec 16, 1966); International Covenant on Economic, Social, and Cultural Rights (Dec 16, 1966); Convention on the Elimination of All forms of Discrimination Against Women (Dec 18, 1979); Convention against Torture and Other Cruel, Inhuman or Degrading Treatment or Punishment (Dec 10, 1984); Convention on the Rights of the Child (Nov 20, 1989); International Convention on the Protection of the Rights of All Migrant

United Nations has explicitly endorsed the importance of individual human rights in the development and promulgation of the RtoP doctrine.[3]

Even accepting the existence of such social facts, one might remain skeptical of the moral standing of individuals as members of the international moral community. For example, one could argue that such conventional facts about the legal practices and rhetorical devices employed fail to provide us with a ***moral*** justification. In other words, echoing the legal positivist's distinction between morality and legality, the skeptic might argue that there is a category difference between social facts and moral facts. The skeptic could likely muster an extensive list of social facts that point in the opposite direction.[4] In the end, even though international law and political rhetoric provide some evidence of the standing of individuals as members of the international moral community, such social facts are not dispositive.

For these reasons, a ***moral*** justification requires a free-standing moral argument in support of the proposition that individuals should be recognized as having standing within the international moral community. There are two arguments that weigh in favor of such recognition. The first is based on the proposition that understanding the moral nature of social and political institutions should be grounded in a commitment to value individualism.[5] The second is based in a demonstration that a rejection of the proposition that individuals have moral standing within the international moral community leads to morally absurd outcomes.

As to the former, it is important to briefly explain what I mean by "value individualism." Simply put, this means that our understanding of moral value within the political space—local, domestic, transnational or international—begins with a commitment to the proposition that individuals are the ultimate source of moral value. In other words, respecting the moral worth of individuals is a foundational ethical requirement of any social or political institution. Though one might be concerned that this is too strong an assumption and in need of further defense, it is my contention that any moral theory that denies the moral worth of individuals leads to the sort of absurd results identified in the second argument. As such, and though the two arguments are distinct, they are mutually supportive.

This should not be confused with the more specific, and stronger, claim that value individualism implies that one is never justified in harming an individual or interfering with an individual's interests or autonomy. All that respect for value individualism requires is that the moral worth of individuals is respected, and that social and political institutions treat individuals as appropriate objects of moral concern. This weaker understanding allows for states to burden or even harm an individual if doing so can be morally justified but doing so would require a justification that includes

---

Workers and Members of Their Families (Dec 18, 1990); and, International Convention for the Protection of All Persons from Enforced Disappearance (Dec 20, 2006).

[3]International Commission on Intervention and State Sovereignty, Summary Report (2001). See also United Nations World Summit Outcome Document (2005).

[4]Goldsmith and Posner.

[5]Altman and Wellman (2009), pp. 41–42.

more than simply weighing moral reasons for and against the action or policy in question.

In my considered judgment, respect for value individualism is best served by the ascription of individual rights. This assertion is based on the claim that a part of what makes individuals the ultimate source of value within the political space is our capacity for moral autonomy—our ability to be self-directed and morally motivated. Consequently, respecting the value of an individual in an environment where they could be dominated by other actors requires that one respect their autonomy; and, since rights serve to protect individuals from the threat of such domination, this is best served by recognizing individuals as rights-holders. This does not, however, mean that rights are absolute moral claims. As Henry Shue notes, even basic rights are limited by what is reasonable to expect or demand.[6] Rights alter the moral calculus in our practical deliberations, but they don't supplant or supersede such deliberations.

Many may want to see a more detailed defense of the claim that respect for value individualism is best served by individual rights. Doing so, however, would take this discussion too far afield, and it is unnecessary since a commitment to value individualism can still generate duties borne by states and owed to individuals without the need for rights. A commitment to value individualism merely requires that states refrain from acting in ways that fail to demonstrate respect for the moral worth of individuals. Consequently, even if it were the case that individuals are not rights-holders, states would still owe duties to individuals.

In the end, whether the duties owed are correlative to rights held by individuals or are primary duties generated directly by the demand that social and political institutions treat individuals as proper objects of moral concern, they burden states with a moral duty to respect the moral value of individuals. This would, at the very least, include duties not to harm individuals and duties not to use individuals as mere means, including duties owed to distant others.

One might concede that states have moral duties not to harm their own citizens but argue that the claim that states have duties not to harm distant others is more controversial. Though a thorough defense of this proposition is not possible here, the following should suffice for present purposes. Social and political institutions are artefacts of our creation and are intended to serve our moral and practical interests, but they are also limited. They are unable to gain a moral power that we, as individuals, lack. Much as we are limited by the duties we have to refrain from harming others, so too are the social and political institutions that mediate our interactions with others, including distant others.

The second moral argument in defense of the proposition that individuals ought to be recognized as having moral standing within the international moral community begins with a counterfactual. Imagine a situation in which the normative framework of international relations ***does not*** ascribe any rights to individuals with correlative moral duties borne by states ***nor*** that states bear any primary duties owed to

[6]Shue (1996).

individuals, including the duty not to harm. The implications of this assumption are morally absurd. Under this hypothetical, states are morally independent of the individuals they are intended to serve.

By implication, states would be unencumbered by any moral requirement to act in ways that treat individuals as appropriate objects of moral concern. Consequently, if it were the case that states need not treat individuals as objects of moral concern, then were a state to act in ways or pursue policies that harmed every individual affected by its action, it would not have wronged them. The moral absurdity of this counterfactual should be apparent.

This moral absurdity does not depend on jurisdiction. It would be no less absurd to claim that social and political organizations need not treat distant individuals as objects of moral concern. Individuals are objects of moral concern and every social and political institution should be understood to have moral duties to treat individuals as such. And, as noted, treating individuals as objects of moral concern, at the very least, requires that the social and political institutions that govern our lives ought not harm, or contribute to policies or practices that harm individuals, at least not without a compelling moral justification. These arguments are consistent with and supported by the discussion in Chap. 7 of the nature of the harms caused and moral imperatives generated by climate change.

8.3 The Content of Rights Held and Duties Owed

Having established that both states and individuals have standing within the international moral community, the next step to assessing whether climate change falls within the legitimate zone of sovereign discretion requires that we flesh out the content of such rights and duties. The challenge of climate change is not within the limits of sovereign discretion inherent to the notion of sovereignty if climate change and its effects involve a failure to respect such rights or fulfill the duties owed to other members of the international moral community.

8.3.1 States

As to states, states have rights to territorial integrity, political independence and sovereign equality. Collectively, these rights imply a broader principle of non-intervention, a right to be free from the interference of other states in the domestic affairs of the right-holding state. Correlative to those rights are duties to refrain from interfering with each rights-holding state's enjoyment of the substance of those rights.

Assessing whether climate change constitutes a violation of the rights held by, or failure to fulfill the duties owed to, states, requires first that we determine whether the effects of climate change are at least a *prima facie* violation of such rights or failure to fulfill such duties. Relying on Henry Shue's understanding of rights as being entitlements to the enjoyment of the substance of a right,[7] the determinative question is whether the effects of climate change interfere with ***any*** state's enjoyment of the substance of its rights.

Thus, determining whether climate change constitutes a violation of a state's rights is about whether the effects of climate change constitute a failure to provide social guarantees against standard threats to each rights-holding state's ability to enjoy the substance of its rights.[8] To answer this inquiry it is necessary that we look to our understanding of what is required for a threat to the enjoyment of the substance of a right to count as a standard threat,

> a standard threat to the substance of a right is one against which protection can be provided, it must be known or knowable, the relationship between the threat and the substance of the right must be direct and either intended to prevent the enjoyment of the substance of the right or with a high probability of preventing such enjoyment, and the threat must not be morally justified.[9]

From this understanding we can identify four characteristics of a standard threat:

1. The threat must, in actual fact or high likelihood, interfere with a right-holder's ability to enjoy the substance of the right in question;
2. The threat must be one of which we are (or should be) aware. It is not a standard threat that we can reasonably expect a duty-bearer to provide protections against if they can't know about the threat or their contribution to it;
3. It must be possible to do something in response to the threat. This includes efforts to stop contributing to or causing the threat, protecting against the threat, and/or preventing others from contributing to or causing the threat; and
4. The threat itself must not be morally justified.

The fourth condition merits further explanation. The justification at issue is not an all-things-considered justification, rather it is about the nature of the threat itself. It must be a threat that, by its nature, is justified. For example, all other things being equal, presenting an existential threat to another person is a standard threat to the victim's enjoyment of their right to physical security. Such threats are not always unjustified. There are even types of killings that are justified. The justification is not, however, on the balance of reasons, rather it is based on the moral nature of the killing. Killing in self-defense is justified; consequently, the justified use of lethal force in self-defense is not a standard threat to the right to physical security of an aggressor.

From this brief discussion of a standard threat we can also draw important implications about when a duty-bearing agent has engaged in a morally culpable act or omission. Assuming that the ability of a rights-holding agent to enjoy the substance of its right(s) is being or has been interfered with, a duty-bearing agent that

---

[7] *Ibid.*

[8] *Ibid.*

[9] Kassner (2012).

is able to provide the requisite social guarantees but fails to act or continues to contribute to the interference, is morally culpable for that failure. In the present circumstances, such failures constitute a culpable moral failure on the part of states.

Whether the specific threats posed by climate change are standard threats begins with an assessment of whether the threat identified interferes with any right-holder's ability to enjoy the substance of its rights. The importance of this question should not be overlooked and is linked to the fact that a threat that does not interfere with a state's ability to enjoy the substance of its rights is morally irrelevant, at least in terms of the implications it has for the practical deliberations of the threat-posing state. One implication of sovereign discretion is that a sovereign state is able to act in ways that others may find morally problematic provided that those actions don't involve a failure to respect the rights of, or fail to fulfill the duties owed to, others. The action may be worthy of moral criticism, but if it is not a violation of a right or failure to fulfill a duty, then it is likely that a state has discretion over it.

One may wonder why all that is required is interference with ***any*** single state's enjoyment of the substance of its rights, contending that a more systemic failure ought to be required if such failures are to identify the limits to sovereign discretion. There are at least three reasons why one should not be moved by this concern. First, the question of whether, ***all-things-considered***, sovereign discretion should be extended over matters that involve the violation of another state's rights, is a matter to be taken up when we consider the instrumental argument for sovereign discretion and the limitations it implies.

Second, each sovereign state's right to sovereign equality requires that even a single violation be accorded its full moral force. A failure to recognize a violation as limiting the sovereign discretion of an offending state is an abandonment of the of the right held by each state to sovereign equality, treating the offending state as superior to the victim.

Third, and finally, as previously argued, a single violation is sufficient to place a matter outside of the legitimate zone of sovereign discretion inherent to the notion of sovereignty because a violation of another's rights is, under the argument being pursued, by its very nature unjustified. Thus, it is beyond the discretion of a duty-bearing state. It is worth noting that due to the nature of the harms of climate change, as a practical matter, there are far more potential violations than is necessary to establish a *prima facie* case of violation.

Though not a comprehensive list, the examples discussed in Chap. 7 should suffice to move our analysis forward. First, climate change is causing sea levels to rise.[10] There is perhaps no clearer instance of interference with the rights of a sovereign state than the existential threat rising seas pose to the territorial integrity of a state. The very ability of a state to enjoy the substance of its sovereign rights is grounded in its territorial nature and rising sea levels pose a threat to the existence of the territory that defines a state.

---

[10]Intergovernmental Panel on Climate Change, *Fifth Assessment Report* (2014).

Also, coastal (and tidal) communities around the globe are facing a future of flooding, loss of potable water and arable land, amongst other threats posed by rising seas. Here again, we see numerous sovereign rights implicated. The fact that rising sea levels will change the habitable area of many states is a clear interference with those states' enjoyment of their right to territorial integrity. Additionally, how a state governs its internal political affairs will be impacted by the need to address rising seas. States will be compelled by man-made circumstances beyond their control to employ their resources and pursue policies that are foisted upon them by the actions of other states. Their political independence will be negatively impacted.

In addition, rising sea levels pose a significant threat to numerous coastal ecosystems. Such threats constitute both an interference with the rights held by sovereign states and an interference with the international community's shared environmental heritage. As to the former, enjoying the substance of the right to territorial integrity means that a right-holding state has (normative and actual) control over the natural resources within its territory, including its environmental resources. The threat to coastal ecosystems posed by climate change constitutes an interference with the ability of states to enjoy the substance of their right to political independence as well, because the choice of how to manage their environmental and natural resources is being interfered with by forces beyond their control.

As to the proposition that rising sea levels and the destruction of coastal ecosystems constitutes a morally relevant interference with the international community's shared environmental heritage, if we start with the proposition that the various ecosystems of the planet are interconnected and interdependent, then harm to any one ecosystem will have impacts that reach far beyond the boundaries of a single sovereign state. Consequently, when such fragile ecosystems are impacted, there is (at least) a *prima facie* case of interference with an interest—if not a right—that the international community has in the protection of, and harms caused to, the coastal ecosystems under threat.

In the end, there seems to be little reason to doubt that rising sea levels interfere with the ability of many states to enjoy the substance of their sovereign rights. Thus, having established that rising sea levels constitute a *prima facie* case of interference with the ability of states to enjoy the substance of their rights, we must now answer whether the other characteristics of a standard threat can likewise be accounted for.

Whether social guarantees against the threats posed by rising seas can be provided (3.) depends on our ability to know of the threats and their origins (2.). Consequently, it makes sense to determine whether we know (or should know) about the threat posed and their cause(s) well enough to adequately assess whether the required social guarantees can be provided. One aspect of this epistemic hurdle is easily overcome. Namely, whether we know of the threat rising sea levels pose to the ability of states to enjoy the substance of their rights, the answer—based on the prior discussion—seems incontrovertible.

There may, however, be some disagreement over when the connection between climate change, rising sea levels and the threats posed were or should have been known. Though making that determination is relevant to assessing moral culpability, it is not determinative of whether we know or should know about the underlying threat posed to the ability of a right-holder to enjoy the substance of its rights. Rising sea levels ***are*** interfering with the ability of states to enjoy the substance of their rights.

When states knew or should have known about the threat rising sea levels posed to the ability of states to enjoy the substance of their rights is a matter over which reasonable people can disagree. This is a consideration that will matter most when we turn our attention to the design of institutions intended to address the challenge of climate change. Addressing such matters in legitimate ways is a task that a well-designed and morally defensible institution must accommodate.

Nonetheless, knowledge of the threat rising sea levels pose to the ability of a right-holder to enjoy the substance of its rights is not enough. We must also be able to determine, with a sufficient degree of specificity, what is causing or contributing to rising sea levels. It is only when we have that understanding that we can determine whether and how the requisite social guarantees can be provided. With that said, the cause of rising sea levels is climate change, and climate change is caused by the acts and omissions of humans.

Next, we must determine whether there are there social guarantees that can be provided to protect against the threats posed by rising sea levels? Each of the causes identified are within our control. Better policy choices, regulations, economic incentives, etc. could be employed to provide the requisite social guarantees against the threat rising sea levels pose to the ability of states to enjoy the substance of their rights.

Revisiting an earlier objection, one might argue that there is little we can now do to stop rising seas. Consequently, there is little that can be done to provide social guarantees against the threats posed by rising sea levels. This objection is based on the proposition that by failing to provide the requisite social guarantees against standard threats to the enjoyment of the substance of a right we have created a state of affairs in which the harms have already occurred, and thus can't be prevented. In short, the objection is based on it being the case that a failure to provide the requisite social guarantees in the past serves as a *post hoc* excuse now.

On its face, this objection has rather absurd implications for moral culpability and responsibility. Additionally, the objection is based on a misunderstanding of what it means to provide social guarantees to the enjoyment of the substance of a right. Providing social guarantees against standard threats is a diachronic notion. It is not just about preventing present interference against standard threats to the enjoyment of the substance of a right, it is also about rectifying past failures and preventing future ones.

Consequently, failing to prevent an interference with the ability of a right-holder to enjoy the substance of a right does not absolve a correlative duty-bearer of their responsibility. Rather, what is called for is rectification for past failures, termination of those activities that continue to cause or are contributing to the threats in question, efforts to mitigate the impact of the threats being faced by the aggrieved right-holders, support for the aggrieved parties in their efforts to adapt in the face of such threats, and actions taken and/or institutions and practices created to ensure against future interference.

As to whether states can provide the necessary social guarantees, the causes of the harms of climate change themselves provide valuable insights into what social guarantees can be provided. At the very least, the activities that states are engaged in and policies being pursued that contribute to climate change can, and should be, stopped. In addition, states should incentivize more sustainable economic development, cleaner fuels and less energy intensive lifestyles. Efforts should be made to mitigate the extent to which any state's ability to enjoy the substance of its rights is interfered with by rising sea levels and support should be provided for states facing the need to adapt. Finally, since there are also issues of historic and distributive injustice that dictate that the developed world assist the developing world in their efforts to combat climate change.

There may be a temptation to discuss in detail the specific responses to challenge of climate change that might be called for. The point here, however, is not to identify what specific policies should be pursued. Rather, what is needed is a demonstration that the provision of social guarantees is possible. On this point, the answer is clear, whether the policies pursued are focused on mitigation, adaptation or a mix of the two, the provision of social guarantees is possible. As to the question of what specific policies should be pursued, it is a matter that is subject to reasonable disagreement, and consequently, is a political matter to be addressed by the institutions responsible for responding to climate change. A more specific discussion of that issue is taken up in Chaps. 10 and 11.

Three of the four characteristics of a standard threat have now been established. The last to be discussed is whether the threats rising sea levels pose to the ability of states to enjoy the substance of their rights are justified. If not, then the failure of states and the international community to provide the requisite social guarantees against standard threats constitutes a violation of the rights of the aggrieved states.

As noted, to be justified it is not enough that the balance of reasons provides us with an all-things-considered reason for failing to provide rights-holding states with the requisite social guarantees to the enjoyment of the substance of their rights. Rather, to be justified, it must be the case that there is no duty owed, that there is no moral relationship in the circumstances under consideration capable of generating countervailing moral burdens. In other words, it is not a case in which a violation of rights is permitted as the least bad option, but that there is no violation.

Whether the failure to provide the social guarantees necessary to ensure that states can enjoy the substance of their rights can be justified, depends on whether, under the circumstances, a right-holding state is entitled to enjoy the substance of its rights, whether it is entitled to such social guarantees. If not, then the failure to provide the requisite social guarantees can be justified. Conversely, if a right(s)-holding state is entitled to enjoy the substance of its right(s), and thus entitled to the social guarantees necessary to make that possible, then a failure to provide such guarantees is not justified.

Thus, we must determine which description best captures the failure on the part of the international community to provide social guarantees against the threats climate change poses to the ability of right(s)-holding states to enjoy the substance of their rights. Are states, when faced with the specific threats posed by climate change, entitled to enjoy the substance of their rights, or is there some compelling moral argument for why they are not so entitled?

We should approach the question from the presumption that a right(s)-holder is entitled to enjoy the substance of its right(s). This is an implication of what it means to have a right. To treat the matter as an open question or to require the right-holder to defend their entitlement would defeat the very purpose of rights, which is to protect certain valuable interests of the right(s)-holder against the capriciousness of others and against events beyond the right(s)-holder's control. Requiring right(s)-holders to defend their entitlement to the substance of a right would be, as a practical matter, equivalent to denying the existence of the right itself. It should be those who are responsible for bearing the correlative duty(ies) to provide the necessary social guarantees, but have failed to do so, that bear the burden of persuasion. The duty-bearers must demonstrate that the right(s)-holder is, under the circumstances being considered, not entitled to enjoy the substance of its rights.

If correct, then our analysis begins by presuming that states are entitled to the social guarantees necessary to ensure that they can enjoy the substance of their rights against the threats posed by climate change. Rebutting this presumption requires more than an argument that demonstrates that in some instances, states might not be so entitled, rather the argument must demonstrate that the presumed entitlement is lacking for all states in every instance of a threat posed by climate change. This follows from the fact that a single violation of a right is sufficient to establish that climate change is outside of the legitimate zone of sovereign discretion.

Thus, the task of those seeking to deny either that climate change is a standard threat or that the failure of the international community to provide the requisite social guarantees against the threats posed by climate change is a morally culpable one is a difficult—if not Sisyphean—task. To rebut the presumption, one must demonstrate that the right does not apply.

An example of how this might work in the case of an individual right should be instructive. Imagine that Candace is shot by Dani. Immediately, we are faced with a question of whether Candace's right to physical security has been violated by Dani. We start with the presumption that Candace is entitled to enjoy the social guarantees necessary to enjoy the substance of her right to physical security and further that the failure to provide those necessary social guarantees in this instance amount to a culpable moral failure on the part of many correlative duty-bearers, especially Dani.

We go on to discover, however, that in the events leading to Dani's use of force, Candace posed an existential threat to Dani, who is a right-holder. Consequently, to avoid harming Candace, Dani would have been required to sacrifice her life. We further discover that Candace's actions were the initial cause of the threat and were not in response to or provoked by any actions taken by Dani. Thus, *ceteris paribus,* Candace was also solely morally culpable for the confrontation and the various threats involved, a paradigm case of self-defense. Here, most would agree that Candace's actions made her liable, that she was not entitled in this instance to social guarantees against the threat to her physical security posed, at least not from Dani. Consequently, Dani bore no correlative duty to provide such social guarantees because the right did not protect Candace in this instance.

What we learn from this example is that determining whether the right in question applies is about the relationship between those whose moral relationship is defined by the right. It is as much about the right(s)-holder not being entitled to the ability to enjoy the substance of the right as it is about the correlative duty-bearer not being required to provide the necessary protections to which the right(s)-holder would normally be entitled. It is about that relationship, not some broader understanding of relevant moral considerations that lie outside of that relationship.

Consequently, the right and its correlative duties do not apply when something has been done to interfere with the moral relationship in question. Specifically, the right(s)-holder must have done something to forfeit their claim to the right in the circumstances at issue, which must involve the duty-bearer who has failed to provide the social guarantees normally required by the right, as it is ultimately about whether that duty-bearer's failure to fulfill the correlative moral burden is justified.

Have rights-holding states done anything to forfeit their entitlement to social guarantees against the standard threats posed by climate change? Many, though certainly not all, states have engaged in conduct that is morally culpable. This, however, fails to justify the claim that actions by rights-holding states constitute a forfeiture of their entitlements. First, not all states have engaged in morally culpable behavior, and for that reason there can be no question that they have not forfeited their rights. Second, it is not clear that even those states that have engaged in morally culpable behavior have severed the moral relationship between themselves and correlative duty-bearers. Finally, the right and duty are ongoing. There is as much a need for social guarantees to protect against future threats as there is for the demand that states act to rectify past, and guard against present, harms.

A different objection might be pressed on behalf of those states that have not engaged in morally culpable action, at least not culpable for climate change. To be specific, one might argue that those states that are largely nonculpable for, yet are facing existential threats from, climate change, lack any duty to provide or contribute to the social guarantees necessary for protecting the ability of morally culpable states to enjoy the substance of their rights. Though there are certainly good reasons—moral and otherwise—for recognizing the moral difference between culpable and nonculpable states, the proper response is neither the forfeiture of a culpable state's rights nor the abdication of a nonculpable state's duties.

For one thing, as a practical matter, we can't disaggregate the impact of the protections of such social guarantees, at least not in the case of climate change. Efforts to mitigate the causes of climate change are a public good. A failure to adjust one's behavior to combat a changing climate affects culpable states and nonculpable states alike. There may, however, be specific forms of social guarantees, for example material support for adaptation, that do admit of more selectivity than others. To be specific, culpable states may have a duty to provide aid to less-able nonculpable states to assist in the latter's efforts to respond to challenge of climate change.

In addition, as was noted, the duty to provide social guarantees against standard threats is not merely, or even most importantly, a basis for the rectification for past harms. It is also about future possibilities and the aspiration to ensure that right(s)-holders can enjoy the substance of their right(s) in the future. Consequently, it is not clear that past actions would negate a right(s)-holder's entitlement to the social guarantees directed at protecting their future ability to enjoy the substance of their rights.

Finally, combatting climate change in a legitimate way requires highly coordinated action by the international community, ***not*** (even if in good faith) efforts by individual states acting in loose confederation. Different degrees of culpability may, however, serve as bases for differentiating between the contributions each state should be expected to provide to support the needed social guarantees around the world. That, however, is not a forfeiture on the part of right(s)-holders or a justification for failure on the part of duty-bearers. How such responsibilities are allocated is best addressed, like other political questions, through legitimate political institutions.

In the end, we can conclude that duty-bearing states and the international community are not justified in their failure to provide the social guarantees necessary to protect the entitlement that states have to enjoy the substance of rights threatened by climate change. What this means is that the failure to provide those necessary social guarantees is a morally culpable failure. Thus, the failure constitutes a violation of the rights of other sovereign states. Consequently, climate change is not a challenge within the legitimate zone of sovereign discretion.

8.3.2 Individuals as Members of the International Moral Community

As noted, states are not the only relevant members of the international moral community. Yet, due to the fact that we have established that the challenge of climate change lies outside of the legitimate zone of sovereign discretion inherent to the notion of sovereignty, one may be skeptical of the need to assess whether the duties owed to individuals by states have been or are being violated by the effects of climate change. There are at least three reasons for proceeding.

First, I would rather be criticized for providing too much support for the proposition that the challenge of climate change lies outside of the legitimate zone of sovereign discretion rather than too little, to err on the side of over-, rather than under-, determination. Second, and looking ahead to Chap. 9, assessing the instrumental value of including the challenge of climate change within the legitimate zone of sovereign discretion depends on whether expanding the moral discretion afforded sovereign states enhances the international community's ability to legitimately address the challenge of climate change. If the challenge of climate change is constituted, at least in part, by the harms suffered by individuals, then understanding whether the challenge of climate change can be best addressed through the expansion of sovereign discretion depends on understanding the causal and moral relationship between states, the harms of climate change, and the harms suffered by individuals. Third and finally, to complete the transition from theory to practice,

Chap. 11 offers some suggestions for institutional design and reform. Those suggestions are guided by various normative commitments, especially practical effectiveness and moral and political justifiability. Understanding how the actions (and omissions) of states have caused climate change and thus led to the harms being suffered by individuals is essential to understanding whether the suggestions proffered are legitimate.

Whether one thinks the moral standing of individuals within the international moral community is grounded in rights or duties, it is important to assess the extent to which the actions or omissions of states constitute a culpable moral failure. It would be practically impossible to provide an exhaustive list of the harms suffered by individuals caused by climate change; nonetheless, and though such thoroughness may be desirable, it is not necessary. The harms to individuals caused by climate change discussed here are sufficient for assessing whether the actions (or omissions) of states causing such harms constitute a culpable moral failure.

The specific impacts of, and harms caused by, climate change being relied upon here were previously identified in Chap. 7. Rising sea levels, changes to hydrologic cycles and the broader implications for what were once reliable weather patterns are impacting individuals in a myriad of problematic ways. For example, due to climate change, individuals are facing food and water insecurity, loss of property and livelihoods, and interference with the social and political stability of their communities.

Having identified the relevant harms of climate change, to assess whether such harms constitute a violation of the rights held by or a failure to fulfill duties owed to individuals requires that we understand what rights and duties are at issue. This requires that we understand the scope and content of these rights and duties. As to the scope, it must be the case that the rights and duties at issue are constitutive of a moral relationship between states and individuals. As to the content, whether the harms of climate change constitute culpable moral failures, depends on the specific content of the rights and duties in question.

Before moving on, it makes sense to address the possibility that individuals are not rights-holders within the normative framework of international relations—a proposition that I do not endorse. Assuming the truth of this proposition, the alternative is not that, since individuals are not rights-holders, there is no moral relationship between states and individuals, rather the moral relationship between states and the international community on one hand and individuals on the other is defined by primary moral duties that states owe to individuals. At the very least, such duties would include duties not to harm and may also include limited duties to respect the inherent human dignity of individuals – to not treat individuals as mere means.

For brevity sake, I will set aside a more thorough explication of these duties and their implications and will move forward under the assertion that, at least with regards to the duty not to harm, this is sufficiently similar to the correlative duties that states have to respect the basic rights of individuals. As a consequence, if states and the international community have failed to fulfill the duties they bear correlative to basic rights, then they have likewise failed to fulfill primary moral duties they have to refrain from harming individuals. There are no doubt many morally relevant differences between the two accounts, but not practical differences for present purposes.

Notwithstanding this, the most salient difference between our analysis of correlative duties versus such primary duties, is more about how we think about the nature of the justification for a culpable moral failure, than it is about a substantive difference between what constitutes such failures. In the case of duties correlative to a right, the matter of justification is dependent on whether, in the particular circumstances, the rights-holder is (or is not) entitled to the social guarantees that would normally be required to fulfill the duties correlative to the right. On the other hand, determining whether a failure to fulfill a primary duty is justified will ultimately depend on whether the intended beneficiary is entitled to the substantive benefit that the duty-bearer is bound to confer on the beneficiary.

Whether the harms of climate change suffered by individuals are a culpable moral failure, depends on whether the actions of states contributing to climate change constitute a failure to fulfill any duties that states owe to individuals. Despite the fact that, as a practical matter, it doesn't matter whether the duties in question are correlative to rights held by individuals or primary duties borne by states to which individuals are the beneficiaries, the initial analysis that follows is focused on the duties correlative to the basic rights held by individuals.

This rights-based analysis, is again dependent on the work of Henry Shue and the proposition that rights-holders are entitled to have the enjoyment of the substance of their rights protected by social guarantees against standard threats. If we assume that persons have basic rights to physical security and subsistence, then individuals are entitled to have social guarantees provided to protect their ability to enjoy their physical security and to subsist against standard threats, which correlative duty-bearers are bound to provide.

A more explicit statement of what is meant by a "basic right" is in order. Basic rights are, as Shue argues, necessary for all other rights.[11] Thus, it is not their relative moral value that makes them basic but the fact that they are foundational to any system of individual rights that makes them basic.[12] Basic rights are not dependent on political affiliation, nationality or citizenship. Yet, *ceteris paribus,* the state may be best situated to provide the requisite social guarantees; thus, we have good reason for ascribing primary responsibility for fulfilling the correlative duties to states, especially the state within which the individual is a resident or citizen.

States, however, often fail to respect, and throughout history have been some of the worst violators of, basic rights. Consequently, when states do fail, the international community is taken to be a secondary bearer of the correlative duties.[13] There are good reasons for concluding, assuming that individuals have basic rights, that states and the international community bear the correlative moral duties to provide

[11]Shue (1996).

[12]*Ibid.*

[13]See International Commission on Intervention and State Sovereignty, Summary Report (2001).

the social guarantees necessary to ensure that individuals are able to enjoy the substance of their basic rights.

Having explained the concept of a basic right and what it implies, the next question to answer is whether the effects of climate change are interfering with the basic rights of individuals. Though the connection may not be obvious, in the case of the right to physical security, climate change is causing more and more intense severe weather around the globe, putting individuals and their communities in harm's way, resulting in deaths numbering in the tens of thousands annually.[14] As to the right to subsistence, the threat of rising seas and altered hydrologic cycles has caused many to face worsening food and water insecurity,[15] direct threats to the ability of individuals and their communities to subsist—a depressingly low bar.

Finally, whether climate change constitutes a standard threat, and perhaps more importantly, whether the failure to provide the requisite socially guaranteed protections are a morally culpable failure on the part of states, constituting a violation of the basic rights held by individuals, depends on whether states are justified in their failure to provide the necessary social guarantees. Which, in turn, depends on whether, in the circumstances under consideration, the relevant rights-holders are in fact entitled to such protections.

Similar to our understanding of how we ought to think about the rights of and duties owed to states, we should begin with the presumption that individuals are entitled to the social guarantees necessary to protect their enjoyment of the substance of their rights and/or that they are entitled to the benefits that duty-bearers are bound to confer on individuals as beneficiaries. Rebutting this presumption and demonstrating that individuals are not so entitled is, again, a difficult task. One would have to demonstrate that an individual has done something for which they are morally culpable that renders them liable to the threat of harm posed by the presumptive duty-bearer. When it comes to the harms of climate change, it is difficult to imagine what an individual, nevertheless all individuals, could do to forfeit their basic rights or their entitlement to not be harmed, especially when they are acting in ways that are often dictated to them or incentivized by the policy choices states make.

From this discussion of the harms of climate change suffered by individuals, we are inexorably led to the conclusion that such harms constitute a culpable moral failure on the part of states. Specifically, states have failed to fulfill the duties they owe to individuals. Regardless of whether those duties are correlative to the basic rights of individuals or primary duties borne by states not to harm individuals or not to treat individuals as mere means, the contribution that states are making to climate change is a violation of such duties. Consequently, we are yet again led to the conclusion that the challenge of climate change does not fall within the legitimate zone of sovereign discretion.

---

[14]"Climate crisis already causing deaths and childhood stunting, report reveals," *The Guardian* (July 31, 2019).

[15]*Ibid.*

8.4 Conclusion

In the end, there can be little doubt about whether the challenge of climate change is within the limits to the legitimate zone of sovereign discretion inherent to the notion of sovereignty. Whether one is assessing the extent to which the harms caused by climate change constitute a failure to respect the moral standing of states or fulfill the duties owed individuals as members of the international moral community, the answer is the same. The matter is overdetermined.

References

Altman, Andrew, and Christopher Heath Wellman. 2009. *A Liberal Theory of International Justice*, 41–42. Oxford University Press.

Beitz, Charles R. 1999. *Political Theory and International Relations*. Princeton, NJ: Princeton University Press.

Kassner, Joshua J. 2012. *Rwanda and the Moral Obligation of Humanitarian Intervention*. Edinburgh, UK: Edinburgh University Press.

Shue, Henry. 1996. *Basic Rights: Subsistence, Affluence, and U.S. Foreign Policy*. Princeton, NJ: Princeton University Press.

Chapter 9
Applying the Framework: Limits Implied by Instrumental Value of Sovereignty

The fact that climate change does not fall within the legitimate zone of sovereign discretion inherent to the notion of sovereignty itself does not settle the matter. Whether there is an instrumental justification for including the challenge of climate change within the legitimate zone of sovereign discretion remains to be determined, and begins by asking whether including the challenge of climate change within the sovereign discretion of states enhances the international community's ability to respond to climate change.

The instrumental nature of the underlying moral justification for sovereignty, however, makes this more complicated since state sovereignty is justified by the instrumental role it plays in the international community's pursuit of several other morally valuable ends. Consequently, it is not enough to assess the extent to which extending sovereign discretion will enhance (or undermine) the international moral community's ability to respond to climate change alone. Rather, we must also assess the impact our choice will have on the ability of the international community to pursue those other morally valuable ends.

An affirmative answer to the initial inquiry regarding the instrumental relationship between sovereignty and the international community's ability to respond to the challenge of climate change, would end the inquiry, as it would constitute an independent moral justification for extending the legitimate zone of sovereign discretion. In which case there would be no need to assess the impact the extension of sovereign discretion would have on the other morally valuable ends served by sovereignty.

If, however, the answer to the initial inquiry is negative, that the international community's ability to respond to the challenge of climate change is not enhanced by extending sovereign discretion to include the challenge of climate change, then we must also assess the extent to which excluding the challenge of climate change impacts the international community's ability to pursue other morally valuable ends, like political self-determination and international peace and security. In other words, since the moral argument justifying sovereignty is an instrumental one grounded in the role state sovereignty plays in the international community's efforts to pursue

J. J. Kassner, *Climate Change and Sovereignty*, AMINTAPHIL: The Philosophical Foundations of Law and Justice 10, https://doi.org/10.1007/978-3-030-73578-4_9

various morally valuable ends, the assessment of the instrumental value of excluding the challenge of climate change from the legitimate zone of sovereign discretion must account for the impact that choice would have for the ***overall*** instrumental value of state sovereignty.

Thus, answering whether sovereign discretion should be extended for instrumental reasons to include the challenge of climate change unfolds in three steps. First, we must consider whether the international community's ability to respond to climate change is enhanced by extending sovereign discretion to include the challenge of climate change. Second, assuming a negative answer, we must assess the overall impact of not expanding sovereign discretion. Finally, we must assess whether, all-things-considered, including the challenge of climate change within the legitimate zone of sovereign discretion is instrumentally justified.

9.1 Preliminary Matters

Before determining whether sovereign discretion ought to be extended to include the challenge of climate change, there are several preliminary matters that need to be addressed. To begin, there are two potential objections that need to be considered. Both objections reject the salience of inquiring into a possible instrumental justification for expanding sovereign discretion to include a challenge that has already been determined to lie outside of the limits to the legitimate zone of sovereign discretion inherent to the notion of sovereignty itself, arguing that expanding the legitimate zone of sovereign discretion over a matter already determined to be non-discretionary cannot be justified.

On the one hand, one might object that expanding the legitimate zone of sovereign discretion to include climate change is conceptually problematic. Simply put, since we have already determined that the challenge of climate change lies outside of the legitimate zone of sovereign discretion, that limitation is an identification of the limits to the legitimate zone of sovereign discretion. This argument is based on the proposition that rights, and their limitations are not subject to the sort of instrumental reasoning being relied upon here. You can't employ instrumental reasoning to expand or contract rights.

Perhaps even more compelling, at least as an ethical matter, is the argument that the limitations to the legitimate zone of sovereign discretion identified here, in relation to the challenge of climate change, are grounded in the fact that it has already been demonstrated that the effects of climate change constitute moral wrongs—violations of rights held by or failures to fulfill duties owed to those with moral standing as members of the international moral community. Thus, expanding sovereign discretion to include the challenge of climate change would fail to treat those violations and failures with the moral seriousness they deserve.

Though I am sympathetic to these arguments, this discussion is being pursued under an understanding of the moral justification for sovereignty that provides us with a response that, at the very least, makes the question of whether we ought to expand sovereign discretion for instrumental reasons relevant. To clarify, sovereignty and the moral discretion it affords sovereign states is grounded in its instrumental value. Thus, the very rights discussed in Chap. 8 are founded on an interests-based account of rights, wherein the justification for the attribution of rights is extrinsic.[1] Rights are the best ***means*** available to serving valuable interests.

The objection being considered depends on an understanding in which rights are grounded in attributes that have inherent value, like autonomy or liberty. Thus, as the moral justification for sovereignty is grounded in the moral goals pursued by the international community, the equivalence upon which the objection relies is wrong. Sovereign rights are based on, and limited by, an all-things-considered assessment of the moral interests served by sovereignty. From its inception, sovereignty has evolved and adapted to changing times and challenges. Consequently, pursuing the question of extending the legitimate zone of sovereign discretion to include the challenge of climate change is not only conceptually possible, but necessary.

The other objection to be considered is grounded in a similar, seemingly, counterintuitive implication of the choice to expand the legitimate zone of sovereign discretion. Here, however, the underlying concern is a distinctly moral, rather than conceptual, one. Namely, one might argue that there are significant moral risks accompanying the choice to extend the legitimate zone of sovereign discretion beyond the limits inherent to the notion of state sovereignty.

The assumption that climate change is within the legitimate zone of sovereign discretion has led to the moral failures that have placed the challenge of climate change outside the limits to the legitimate zone of sovereign discretion in the first place. In other words, the challenge of climate change lies outside of the legitimate zone of sovereign discretion because it and its harmful effects were caused by states exercising their discretion. It is, at the very least, counterintuitive to contemplate formalizing, and arguably legitimizing, that same discretion.

The question facing us, however, is whether, given the fact that the international community's response, no matter what form it takes, is constrained by the requirement that it be effective. If, all-things-considered, the best institutional option available is to provide states with discretion over climate change, then there is an independent instrumental justification for ascribing that discretion to states. Further, that justification would depend on it being true that the expansion of sovereign discretion would enhance the international community's ability to legitimately respond to climate change.

If we look to human rights as an example, assuming that human rights are matters that lie outside of the legitimate zone of sovereign discretion, then, much like climate change, we start with the presumption that whether and how states respond to human rights concerns is a non-discretionary moral matter not subject to the moral discretion of sovereign states. If, however, the best way to provide, protect and promote human rights is to ascribe the responsibility for doing so to states, this will likely involve granting states ***some*** degree of discretion over how best to provide, protect

---

[1]For a discussion of interests-based rights, see Sumner L. Wayne (1986).

and promote human rights. Such discretion would not, however, be unfettered. Rather, it would be contingent on states using their discretion to provide, protect and promote human rights.

Importantly, the discretion involved is different from the type of discretion implicit in the Westphalian conception of sovereignty, wherein states have the discretion to act in ways that are morally objectionable. The sort of discretion being discussed here, as an intentional and provisional extension of sovereign discretion, is justified because of its contribution to achieving a morally valuable end. As such, it is limited by its effectiveness in that regard. This is as much an evolution of the concept of sovereignty as it is an extension of sovereign discretion.[2]

As a final note, setting aside my own qualms about the underlying argument for the doctrine itself, RtoP involves just such a conditional attribution of sovereign discretion. RtoP is based on the proposition that every individual has human rights and that the best available institution for ensuring that those rights are protected is the state. Consequently, states have been assigned ***primary*** responsibility for the protection of human rights. Along with such responsibility comes a certain amount of discretion over determining the content of such rights and their enforcement. What states do not have discretion over is ***whether*** to protect human rights. Further, should states fail in their responsibility, the international community is the ***secondary*** bearer of this responsibility.

There are at least two elements to the criteria being used to assess whether such an instrumental justification exists that require further explanation. First, we are in need of a more explicit understanding of what is required to determine whether expanding the legitimate zone of sovereign discretion to include the challenge of climate change is justified by its instrumental value. And second, we need a clearer understanding of how to determine the impact not expanding the legitimate zone of sovereign discretion will have on the ability of the international community to pursue other morally valuable ends.

As to the first element, identifying the conditions under which an extension of sovereign discretion for instrumental reasons would be warranted, it is not enough that expanding the legitimate zone of sovereign discretion would neither enhance nor detract from the international community's ability to respond to the challenge of climate change. This follows from the fact that the challenge of climate change conflicts with the limits to sovereign discretion inherent to the notion of sovereignty itself. An affirmative answer to the question of whether sovereign discretion ought to be extended to include the challenge of climate change requires a demonstration of the instrumental value such an extension would have for the international community's ability to respond to the challenge of climate change.

A negative answer to the initial inquiry, leads to the second element. As has been noted, understanding the instrumental value of sovereignty and the choice of whether to extend the legitimate zone of sovereign discretion requires an analysis that goes beyond assessing whether extending the legitimate zone of sovereign

[2]For a related discussion see Shue (2004).

discretion enhances the international community's ability to respond to climate change. The required assessment is complicated by the fact that the instrumental justification for sovereignty requires that one recognize the other instrumental relationships that exist between sovereignty and a range of valuable moral ends pursued by the international community, and that these relationships are not entirely severable from one another. Altering one such relationship will likely have implications for others, and we must account for such impacts if we are to have an adequate understanding of the instrumental value of choosing not to extend the legitimate zone of sovereign discretion to include the challenge of climate change.

Consequently, for the instrumental argument to justify extending the legitimate zone of sovereign discretion to include the challenge of climate change, it must be the case that either including the challenge of climate change enhances the international community's ability to respond, or that failing to do so would, on balance, significantly undermine the other valuable ends being pursued by the international community. In other words, it is not enough that including the challenge of climate change within the limits of sovereign discretion merely keeps the *status quo*, as it has already been demonstrated that the challenge of climate change lies outside of the limits to the legitimate zone of sovereign discretion inherent to the notion of sovereignty itself.

9.2 Assessing the Instrumental Relationship Between Sovereign Discretion and the Challenge of Climate Change

We must now assess whether the zone of sovereign discretion ought to be extended to include the challenge of climate change. To do so, we must first determine the extent to which extending the zone of sovereign discretion will enhance the international community's ability to respond to climate change. Yet, with a challenge as morally and practically complicated as climate change, it would be reasonable to wonder how one could understand the instrumental relationship, or lack thereof, between sovereign discretion and the challenge of climate change well enough to make such an assessment.

Thus, this inquiry must begin with a threshold question. Is it possible that extending sovereign discretion could enhance the international community's ability to respond, in a legitimate way, to climate change? "Possible" here is not a matter of logical possibility, but practicality. It is not enough that in some possible world the expansion of sovereign discretion would enhance the international community's ability to respond to climate change, it must be practically possible in this world. For any who might find this to be philosophically infelicitous, it is important to remember that this is a project seeking to understand the moral nature and limits of sovereignty in our imperfect world.

With that said, there is something about the challenge of climate change and what is needed to respond to it in a legitimate way that makes expanding sovereign discretion practically ineffective as a means to enhancing the international community's ability to respond in a legitimate way. Recall, as a practical matter, a legitimate response to climate change requires a level of coordination antithetical to sovereign discretion. In other words, for the international community to be able to respond to the challenge of climate change in a legitimate way, each state's choices and actions must be highly prescribed.

To clarify, imagine that each state is granted discretion over whether and how to respond to the challenge of climate change, should we expect that the aggregation of individual state's choices and actions will produce an effective response to climate change? This is the world we live in now, and we have ample evidence to conclude that such discretion has not resulted, nor is it likely to result, in an effective and legitimate response to climate change. Unless someone can offer a compelling argument for why we should expect this to change, extending sovereign discretion to include the challenge of climate change will do nothing to enhance the international community's ability to respond to climate change.

As a counterargument, one might look to the RtoP doctrine as a model for how expanding the legitimate zone of sovereign discretion could enhance international community's ability to effectively and legitimately respond to a challenge like climate change. The argument depends on the challenge of human rights being sufficiently similar to the challenge of climate change. More importantly, for the analogy to be appropriate, it must be the case that what is required to respond to the challenge of human rights in a legitimate way is sufficiently similar to what is required to respond to the challenge of climate change in a legitimate way.

Though both are significant challenges facing the international community, they differ in morally and practically significant ways.[3] Focusing first on the nature of the challenges. The challenge of human rights is based in the fact that each of us, as individuals, possess certain rights. Whether one individual's rights are being respected can—at least arguably—be individuated from whether the rights of others are being respected. The challenge of climate change, on the other hand, originates from the collective actions and omissions of various actors around the globe. The causes and effects of climate change cannot be disaggregated in the same way.

This leads to a disanalogy between the required responses. In the case of human rights, it is reasonable to expect that the international community's response to the

[3]It is important to note that this is not meant to imply that human rights and the challenge of climate change are completely unrelated. In fact, the moral challenge of climate change implicates numerous human rights considerations. For example, to the extent that individuals have a right to subsistence and / or political self-determination, climate change contributes to various threats to those rights. I have made the methodological choice to separate the challenge of climate change from the challenge of human rights because an adequate assessment of the implications that the latter has for how we ought to understand the moral nature and limits of state sovereignty would require more than can be accomplished here. With that said, the methodology employed throughout this discussion could be expanded, with some significant effort, to the challenge of human rights.

challenge of human rights could be enhanced by granting states a limited form of discretion over human rights. For example, cultural and historical differences may mean that states are better able to determine—due to their physical proximity and epistemic familiarity—how to respect human rights within their cultures and communities. This does not mean that states have unfettered discretion, but in order to provide, protect and promote human rights in a pluralistic world, this may, in fact, be better than an institutional structure in which states lack any discretion whatsoever.

The same cannot be said of what is required to respond to climate change. As noted, responding to climate change in a legitimate way requires highly coordinated action across political boundaries. Responding to the challenge of human rights in a pluralistic world is enhanced by principles, practices and institutions that can accommodate pluralism while remaining committed to the universal value of human dignity. The challenge of climate change is caused by the global interdependence and interconnectivity of individuals, corporations and political communities around the globe. Thus, an adequate response requires highly structured and thoughtful international coordination, not delegation of individual discretion. Responding to the challenge of climate change requires that differences be resolved, not accommodated.

There is one final point to consider that offers further support for the proposition that extending sovereign discretion fails to enhance the international community's ability to legitimately respond to the challenge of climate change. Recall, in the case of human rights, the sort of discretion being extended is unlike that which defines Westphalian sovereignty. It is ***contingent*** discretion, discretion that is dependent on the bearer of such discretion using it properly.

Looking again to RtoP and human rights, under RtoP the discretion afforded states is arguably effective, thus justifying the conceptual and normative room within which states are free to operate. The same cannot be said in the case of the challenge of climate change. Again, the level of coordination needed implies that the restrictions that would have to be included would make the appropriateness of each state's choices dependent on the actions of other states. There would, *de facto*, be no discretion, even if there was nominal discretion afforded states in principle.

In the end, there is little reason to believe that expanding the legitimate zone of sovereign discretion to include the challenge of climate change would enhance the international community's ability to respond to the challenge of climate change in a legitimate way. In fact, there is compelling evidence to suggest the opposite, that expanding the legitimate zone of sovereign discretion would undermine the international community's ability to respond in a legitimate way. In short, the answer to the initial question of whether there is a direct instrumental relationship between the expansion of sovereign discretion and the international community's ability to respond to the challenge of climate change, is a negative one.

9.3 Possibility of an Overall Instrumental Justification

Having determined that there isn't a direct instrumental justification for extending the legitimate zone of sovereign discretion to include the challenge of climate change, our attention must now turn to assessing the implications not extending the legitimate zone of sovereign discretion will have for the broader instrumental value of sovereignty. We have already discussed how state sovereignty and sovereign discretion are taken to enhance the international community's ability to pursue international peace and security and political self-determination. Now we must assess the extent to which not extending the legitimate zone of sovereign discretion to include the challenge of climate change would impact the relationship between sovereignty and the moral objectives sovereignty is already known to serve.

9.3.1 Climate Change and International Peace and Security

Understanding the impact the choice not to extend the legitimate zone of sovereign discretion will have on the instrumental value of sovereignty starts with an appreciation for the relevant differences between the relationships between state sovereignty and the various morally valuable ends pursued by the international community. Perhaps most important for this discussion is the fact that the instrumental relationship between state sovereignty and international peace and security is an entirely contingent one, whereas the relationship between sovereign discretion and political self-determination is constitutive.

As to the former, international peace and security is not dependent on sovereignty. It is logically possible to have international peace and security without sovereignty. Rather, general respect for sovereignty makes the peace between and security of states more likely. Thus, in relation to the goal of international peace and security, evaluating the impact not expanding the legitimate zone of sovereign discretion will have on this relationship is a matter of assessing the extent to which a choice not to expand the legitimate zone of sovereign discretion interferes with that contingent relationship. To what extent does a choice not to expand the legitimate zone of sovereign discretion to include the challenge of climate change undermine the international community's ability to maintain international peace and security?

One may wonder why this inquiry is relevant at all. Since the relationship between sovereign discretion and international peace and security is conceptually distinguishable from the relationship—or lack thereof—between sovereign discretion and the ability to adequately respond to climate change, what sense is there in exploring the possible impact a choice related to one has on the other. Simply put, the promotion and maintenance of international peace and security is a moral objective worth pursuing. Consequently, if choosing not to expand the legitimate zone of sovereign discretion to include the challenge of climate change were to undermine the international community's ability to promote and maintain international peace and security, that would be a weighty instrumental reason in favor of expanding sovereign discretion to include the challenge of climate change even though doing so could undermine the ability of the international community to adequately respond to climate change.

A thorough assessment of the many ways in which the choice not to expand sovereign discretion could possibly impact the international community's ability to seek out and maintain international peace and security is not possible here. Nonetheless, sinve the underlying justification for the discretion afforded sovereign states is not the value of sovereignty itself, but rather found in the value of international peace and security, much will depend on the relationship between international peace and security and climate change. Further, if it can be demonstrated that addressing climate change enhances the international community's ability to promote and maintain international peace and security, then we should conclude that expanding the legitimate zone of sovereign discretion to include the challenge of climate change cannot be justified by a concern over the impact doing that would have on the ability of the international community to promote and maintain international peace and security.

Thus, we are faced with an empirical question. What is the relationship between climate change and international peace and security? Like many interdependent relationships, this is a complicated one, and our assessment and understanding of it is based on reasonable inferences drawn from incomplete data. Nonetheless, there is a growing consensus that climate change is placing states and regions under tremendous pressure. The Institute for Economics and Peace has identified climate change as a major threat to international peace.[4] And, even the United States' Department of Defense has recognized that climate change is a threat to national security.[5]

As such, whether one is concerned with the impact of existential threats faced by island nations like Tuvalu or the extent to which extreme weather is causing political instability and upheaval in regions around the world, there is ample evidence to indicate that the challenge of climate change is a threat to international peace and security. Consequently, the available evidence indicates that continuing to assume that states have sovereign discretion over whether and how to respond to climate change is less, not more, likely to lead to international peace and security. Ultimately, advancing the cause of international peace and security requires that we respond to climate change, and doing so requires that we not extend the legitimate zone of sovereign discretion to include the challenge of climate change.

---

[4]Sonia Elks, "Climate change seen posing threat to global peace in next 10 years", Reuters (12 June 2019), https://www.reuters.com/article/us-global-climatechange-peace/climate-change-seen-posing-threat-to-global-peace-in-next-10-years-idUSKCN1TD0C6.

[5]"Climate Change Threatens National Security Says Pentagon" (14 October 2014), https://unfccc.int/news/climate-change-threatens-national-security-says-pentagon.

9.3.2 Climate Change and Political Self-determination

We must now assess how not extending the legitimate zone of sovereign discretion would impact the international community's ability to promote and protect political self-determination. To do so we need a more explicit understanding of the relationship between sovereign discretion and political self-determination. The understanding of political self-determination being relied upon here is about the ability of a political community to govern itself free from external interference. Consequently, sovereign discretion is necessary for a state to be politically self-determining.

This understanding of the relationship between sovereignty and political self-determination may lead one to believe that any limitation on the ability of a state to choose how to govern itself, including those that would be related to international efforts to respond to climate change, undermines political self-determination. In which case, the choice not to expand the legitimate zone of sovereign discretion to include the challenge of climate change would conflict with the international community's commitment to political self-determination.

This conflict would only arise, however, under an untenable understanding of the normative breadth of political self-determination. Namely, for it to be the case that political self-determination would, of necessity, be in conflict with a choice not to expand sovereign discretion, it would have to be the case that to be self-governing, what constitutes an internal matter would have to be an extremely broad understanding of internal, one that ignores the transboundary impacts of the policies chosen. As an example, when setting emissions standards, the fact that one state's choices have impacts on neighboring states and the rest of the world would be morally irrelevant.

The near limitless nature of such an understanding of political self-determination is indefensible. For one thing, and as has already been discussed at length, it conflicts with the proposition that states are rights-holding and duty-bearing members of an international moral community, in which the rights of one state are limited by the rights of and duties owed to others. Within the normative framework of international relations, there are limits to what justifiably falls within the normative breadth of political self-determination. A broader understanding would weaken the existing normative framework of international relations and unjustifiably extend political self-determination beyond its moral limits.

Additionally, in a world of greater interdependence and interconnectivity, it is morally problematic for a normative principle to dictate that we ignore the negative externalities of state behavior. The fact that a policy choice or action taken has its immediate effects in a particular locale applied to a particular population does not change the fact that the costs of that choice or action are borne by many outside of the political boundaries of the state making the choice or enacting the policy. Consequently, as a moral matter, there are decisive reasons to be skeptical of the broader understanding of the normative breadth of political self-determination.

In the end, there seems to be little reason to think that the international community's ability to promote and protect political self-determination depends on an extension of the legitimate zone of sovereign discretion to include the challenge of climate change. The choice not to expand the legitimate zone of sovereign discretion only becomes problematic under an unjustifiably broad understanding of the normative breadth of political self-determination.

9.4 Conclusion

From this discussion of the potential for the extension of the legitimate zone of sovereign discretion to be justified by its instrumental value, we can draw several conclusions. Whether one is concerned with the contingent instrumental relationship between sovereignty and international peace and security or the constitutive instrumental relationship between sovereignty and political self-determination, there seems to be little reason to believe that not extending the legitimate zone of sovereign discretion would interfere with the ability of the international community to pursue the valuable moral ends that justify sovereignty and sovereign discretion in the first place. Thus, we can conclude that there is no justification—at least not an instrumental one—for extending the legitimate zone of sovereign discretion to include the challenge of climate change.

It is important to take note of where we stand in the development of the broader project. Having demonstrated that the challenge of climate change lies outside of the limits to the legitimate zone of sovereign discretion inherent to the notion of sovereignty and having now rejected the possibility of an instrumental justification for expanding the legitimate zone of sovereign discretion to include the challenge of climate change, we are justified in concluding that the challenge of climate change does not fall within the legitimate zone of sovereign discretion. Consequently, we can also conclude that the challenge of climate change is outside of the moral limits of state sovereignty.

As such, the application of the framework for assessing the moral limits of state sovereignty to one challenge facing the international community has been completed. A more thorough understanding of those limits will involve an ongoing and ever-evolving process. One that requires a regular evaluation and reevaluation of the extent to which new and emerging challenges facing the international community fall within the legitimate zone of sovereign discretion.

References

Shue, Henry. 2004. Limiting Sovereignty. In *Humanitarian Intervention and International Relations*, ed. Jennifer M. Welsh. Oxford: Oxford University Press.

Sumner, L. Wayne. 1986. *The Moral Foundation of Rights*. Oxford: Oxford University Press.

Chapter 10
Explicating Institutional Legitimacy

Having demonstrated that the challenge of climate change lies outside of the legitimate zone of sovereign discretion, one might be tempted to end this inquiry. Satisfied with the fact that our understanding of the moral nature and limits of state sovereignty have been greatly enhanced. This would be no small accomplishment. Nonetheless, and notwithstanding what has been achieved thus far, the practical aspirations of this inquiry have always been greater. It would be highly unsatisfying to leave our inquiry now as this would fall a few steps short of completing the transition from theory to practice.

As such, in an effort to complete the transition from theory to practice the final two chapters are dedicated to offering some suggestions for institutional design and reform. It is in offering such suggestions for institutional design and reform that the impact of the conclusions reached thus far have their clearest ethical and practical implications. If it had turned out that the challenge of climate change falls (or ought to fall) within the legitimate zone of sovereign discretion, then to be legitimate any institutional design would be constrained by sovereignty. However, due to the fact that the challenge of climate change lies outside of the legitimate zone of sovereign discretion, we are not so constrained; rather, when thinking about institutional design and reform we have a freer hand.

Thus, in in this chapter I develop a set of standards for evaluating the legitimacy of institutions intended to respond to the challenge of climate change. Then in the next, concluding, chapter, I employ that standard of institutional legitimacy to critically assess, what I take to be the most prominent political mechanism currently tasked with addressing climate change, the Paris Climate Agreement. And, finally, I offer some more tangible suggestions for institutional design and reform. Admittedly, for both pragmatic reasons related to the practical limits of this particular project as well as my own lack of expertise in the area of institutional design, I have chosen to refrain from offering any highly detailed suggestions for how an institution tasked with responding to the challenge of climate change ought to be designed, though I do sketch a rough framework for such an institution in Chap. 11.

J. J. Kassner, *Climate Change and Sovereignty*, AMINTAPHIL: The Philosophical Foundations of Law and Justice 10, https://doi.org/10.1007/978-3-030-73578-4_10

10.1 Introducing a Conception of Legitimacy

As noted, the goal of these last two chapters is to offer some suggestions for institutional design and reform. This is as much about identifying standards of legitimacy as it is about the design of legitimate institutions. In this chapter the focus is on identifying the criteria by which we can evaluate whether an institution is legitimate. We have a general and intuitive understanding of legitimacy—at least as the term is used colloquially. As such, one might wonder why any further explanation is needed. Simply put, the concept of legitimacy can be used in a variety of different ways, and a failure to be clear about what one means is likely to cause unnecessary confusion.

An example may help to illuminate the concern. Assuming that, as a matter of commonsense morality, criminal justice systems ought to, all other things being equal, hold individuals accountable for their criminal wrongdoing, a criminal justice system that in some way contradicted this basic moral commitment would be, at least to a degree, morally illegitimate. However, as a matter of juridical fact, nearly all criminal justice systems incorporate technicalities that could—when exploited—undermine the commitment to holding individuals accountable for criminal wrongdoing. In which case, the use of such technicalities may be at the same time morally illegitimate and legally legitimate.

Further, if we follow the notion of legitimacy advocated for by Fabienne Peter, then to be legitimate is to be deserving of respect.[1] Though Peter's work was focused on democratic political legitimacy, a more general lesson can be learned. Namely, the nature of the institution, principle or practice being evaluated dictates the type of respect deserved and, consequently, the type of legitimacy involved. In other words, legitimacy is relative to the object of evaluation. Juridical legitimacy is different from moral legitimacy is different from—what could be called—practical legitimacy.

This should help one understand the significance of the need to state, with a sufficient degree of specificity, what is meant by legitimacy. To begin, the notion being relied upon here is not one to be found in the work of others, and this should not be surprising. As has been noted throughout this discussion, the challenge of climate change is a complicated one. It is, at the same time, a scientific challenge, a political challenge, an economic challenge, a social challenge and a moral challenge. Though it would be unreasonable for us to expect that any institution would be able to address, without compromise, every type of concern that arises with the challenge of climate change, to be legitimate it should be able to respond to them in a way that deserves our respect.

In this discussion, explicating what counts as a legitimate institution serves as the identification of the conditions that are determinative of the extent to which an institution is legitimate. Implicit in this approach is the fact that the conception of

[1]Peter (2011).

legitimacy being developed allows for assessments that admit of degree. One possible institution could be more or less legitimate than another.

One might object that treating legitimacy as a property that admits of degrees is problematic, arguing that legitimacy should be understood as a threshold condition. If an institution passes a certain normative threshold, then it should be considered legitimate. For a variety of reasons, I find this understanding of legitimacy to be unhelpful. For one thing, this is not an inquiry in ideal theory. Consequently, it is possible that whatever institution or reform might be politically possible will fall short of our standard of legitimacy. Not all institutions or reforms are going to fail in equal or equally problematic ways. Some may be ***more*** legitimate than others. Having a standard of legitimacy that admits of degrees will allow for us to assess the relative value of different alternatives.

Additionally, and also related to the non-ideal methodology being pursued, the challenge of climate change is an intractably complicated matter. Consequently, the standard of legitimacy is, and needs to be, complicated as well. This further implies that, in the non-ideal world, any institution or reform we are likely to pursue will involve compromises over many of the substantive areas defining the challenge of climate change, thus making it even less likely that we will be able to create institutions that fully satisfy our account of legitimacy. Again, we benefit by having an account of legitimacy that admits of degrees.

We can now turn to the identification and overview of the features of the conception of legitimacy being discussed here. To begin, to attribute the property of legitimacy to an institution is to claim that the institution deserves respect. The nature of the respect deserved is inexorably linked to the nature of the matters with which the institution is concerned. It is this fact about the linkage between appropriate attributions of legitimacy and the substantive concerns of the institution in question that informs our understanding of the notion of legitimacy being relied upon here.

To be specific, as noted, the challenge of climate change is complicated; consequently, our understanding of what constitutes a legitimate institution must begin with an explicit identification of the various elements that inform our understanding of the challenge of climate change so complex. At the most general level, the challenge of climate change is both a practical as well as a moral matter. As such, what constitutes a legitimate institution must be one that addresses both elements.

This is, however, not specific enough. We need a more detailed understanding of the elements involved if we are to have an account of legitimacy capable of differentiating between alternative institutions. First, and perhaps most apparent, the challenge of climate change is a scientific and technical challenge. As such, a legitimate institution must be, at the very least, capable of understanding and evaluating the current scientific understanding of the challenge and the responses available.

Second, the challenge is a morally complicated one. It is not just about the moral implications that follow directly from climate change and its effects, rather it also must address other relevant moral considerations. For example, a legitimate institution would have to address matters of historic and distributive justice, as well as considerations tied to life in social and political community.

Third, it is a political challenge. Even assuming that we are able to identify with a sufficient degree of specificity a range of possible responses that are morally reasonable and scientifically acceptable, there may be significant disagreements over which of the available responses ought to be pursued and there may be no clearly correct answer to that question. It is a political question about which response ought to be pursued.

Finally, and relatedly, the challenge is an economic one. The choice between pursuing some responses over others will involve trade-offs. Resources will have to be expended and there will be opportunity costs that accompany each choice. It is essential that such choices are made with as much thoughtfulness and intention as is practically possible.

In the end then, our understanding of legitimacy must be able to account for the fact that the challenge of climate change is a scientific, moral, political and economic one. Some may find this delineation of the elements of the understanding of legitimacy either over- or underinclusive. For example, as to the latter, one might argue that cultural and historical elements have a role to play as well; or, as to the former, that in the end all of the elements identified only become relevant if they can be understood to implicate some moral concern related to the challenge of climate change. Regardless, however, of how one wishes to characterize the standard and its elements, one is more likely to create a legitimate institution that is deserving of our respect if they keep this complexity in mind.

There are three final points that warrant mention before returning to the more explicit discussion of the account of legitimacy relied upon in the discussion that follows. First, I take it to have been established that the challenge of climate change is global. Second, identifying the general nature of the elements is not enough, a more explicit statement is needed. This is the primary task of the next section. And, third, it is also important to recognize that though it is possible to think about the challenge of climate change from each individual element's perspective, it is not possible to think about institutions that are intended to respond to climate change in such an isolated manner. Any institution tasked with responding to the challenge of climate change must be integrated across the relevant elements identified.

10.2 Standard of Institutional Legitimacy

We can begin with the general understanding that to be legitimate an institution must be able to accommodate the fact that the challenge of climate change is a matter of scientific, moral, political and economic concern. It is not enough to construct an institution that merely addresses one of these concerns, nor is it enough to construct an institution that simply aggregates, into a single institutional framework, different principles and practices that address each element separately. Rather, as noted, a legitimate institution must integrate consideration of these elements into a unified whole.

With that said, let us begin by recognizing that a legitimate institution must address both the practical and the moral aspects of climate change. This is a conceptual distinction and is being recognized here largely for theoretical clarity. In reality, the two categories overlap in numerous ways. For example, one of the defining features of what it means for an institution to be legitimate from a practical point of view is that the institution in question must be practically effective, but such practical effectiveness is also an essential element of the instrumental value of the institution. To be specific, the instrumental value of any legitimate institution is a matter of great moral concern and is inexorably linked to the practical effectiveness of an institution.

10.3 Practical Effectiveness

With this caveat in mind, explicating the conception of institutional legitimacy relied upon here to evaluate the legitimacy of institutions intended to respond to the challenge of climate change, begins with a discussion of the practical elements of institutional legitimacy. I am beginning with the practical elements of institutional legitimacy because they are, in my considered judgement, more easily understood. Thus, to being, as a general proposition, to be legitimate from a practical point of view, the institution must be practically effective. An institution must make a difference in the international community's ability to respond to the challenge—in this case climate change.

This general understanding lack the necessary specificity to be of much use. More detail is needed. First, it is not enough that the institution would be practically effective under ideal conditions, ones in which those whose behavior is governed by the institution are assumed to adhere unfailingly to the prescriptions promulgated by the institution, to set aside their own interests to follow those prescriptions, and do so in circumstances in which there is clarity and agreement over all of the considerations relevant to the practical deliberations of those whose actions are governed by the institution in question. Rather, what is needed is an institution that is capable of being practically effective in non-ideal circumstances, in a world with all of the competing considerations, ambiguity, and possibility for free-riding and defection that exist in the world we inhabit.

Notwithstanding this commitment to provide an analysis grounded in the non-ideal nature of the world around us, our efforts to create or reform institutions to respond to the challenge of climate change should, as a practical matter, be aspirational. One of the primary objectives has been to lay out a path to a morally better world, but to do so in a practically effective way is to be constrained by what is, to borrow from Rawls, practically politically possible.[2] In other words, though our aim should be guided by higher moral aspirations, institutional design must be grounded in what can be achieved from where we stand right now.[3]

With this in mind, what should our standard of practical effectiveness be in a world of sovereign states? It is one thing to argue that a challenge like climate change is outside of the moral limits of sovereignty and sovereign discretion, it is another to motivate states to do what they ought to do to respond to the problem in accordance with the prescriptions of an institution outside of their control.

It would be unreasonable to expect that states would simply choose to abdicate their decision-making power to a transnational institution, delegating their deliberative powers to an external governing body. Under ideal circumstances, this may be what is called for, but in the world we live in, it is a fantasy to expect states to be so motivated, regardless of the fact that the challenge of climate change is one that every state faces.

At the very least, to be practically effective, what is needed is an institution that makes a difference in the practical deliberations of states, an institution that can change the way a state thinks about the prescriptions generated by the institution and the role such prescriptions ought to play in its practical deliberations. Importantly, even though the institution may not be able to incentive proper behavior through direct enforcement mechanisms, it is equally true that, due to the fact that the challenge of climate change lies outside of the moral limits of state sovereignty, the institution need not respect the sovereignty of states. Thus, making explicit judgments and condemnations, and perhaps sanctions and exclusion, morally permissible.

There is one final point, on this particular criterion of practical effectiveness that needs to be recognized. It is not enough that an institution espouses lofty ambitions and goals and establishes advisory groups and councils that collect information and share it. Such functions are valuable, but they are unlikely to change the practical deliberations of states. In fact, without more, they may do more harm than good, giving the appearance of activity in response to a grave crisis without having much impact on the crisis itself.

Another practical concern that warrants our attention specific to challenges like climate change, is related to the fact that our understanding of the challenge and the effectiveness of our response is heavily dependent on getting certain empirical matters right. First, there are scientific and technological concerns that must be recognized and accommodated. One is relatively straightforward. To be practically effective, an institution must have the epistemic expertise to understand the complicated causal mechanisms underlying climate change and its effects.

[2]Rawls (1971).

[3]It should be noted that this is not meant as an endorsement of the Rawlsian top-down approach to theorizing about justice from an abstract perspective but is meant as a commitment to the proposition that any theorizing about justice should be concerned with what is possible in the world in which we live. I want to thank the anonymous reviewer for alerting me to the possible misunderstanding that invoking Rawls might engender.

In addition, to be practically effective, an institution must be able to distinguish scientifically and technologically feasible responses from those that are not. To clarify, a practically effective institution would be one that relies on its epistemic expertise to narrow the range of responses to those are both scientifically valid and technologically feasible. This should not be confused with the claim that to be practically effective an institution must be able to identify ***the*** correct response. What, in the end, is ***the*** correct response will likely be a matter of reasonable disagreement. What is required is that the various incompatible alternatives underlying such disagreement all pass some minimal threshold of scientific and technological validity and feasibility.

This should not be taken as a reason to exclude the possibility that different alternative responses will be distinguishable based on matters of scientific and technological validity and feasibility. It might be the case that a response is grounded in valid science, but is not technologically feasible, or that the technology exists but that there is a risk that what it is being used for is not supported by a scientific consensus.

Finally, another criterion upon which the practical effectiveness of an institution depends, is related to the complicated nature of the economic considerations implicated by competing alternative responses. This is, again, an epistemic matter. We need to understand the economic impact of various alternative responses if we are to make informed decisions about how to respond to climate change. This does not merely require an estimate of the cost of the response itself, but also of the broader economic implications of the alternative responses available, including possible benefits, opportunity costs and negative externalities.

It is worth reiterating that like the requirement that the institution must enable us to distinguish between scientifically and technologically valid and feasible response from those that are not, this does not mean that, to be practically effective, an institution must be able to identify ***the*** economically optimal alternative. Rather, all that is required is that it must provide us with an understanding of the economic implications of the various responses available so that informed political choices can be made.

10.4 Moral and Political Justification

Shifting our focus to a discussion of what is required for an institution to satisfy the moral element of our understanding of institutional legitimacy, as one will quickly surmise, the discussion of the moral element is heavily dependent on our understanding of what it means for an institution to be practically effective. It should be noted, however, that this is not solely a relationship of dependence of the moral on the practical; rather, it is a relationship of interdependence. What is required for an institution to be morally and politically justified, informs and limits what is required as a practical matter. And, as is discussed below, the information produced—

scientific, economic, technological, etc.—has a significant impact on the moral and political justifiability of the institution and its actions.

Perhaps an even clearer connection between the requirement that the institution be practically effective and the requirement that the institution be morally justified has to do with the fact that the challenge of climate change is not one in which doing nothing is a morally viable option. To be morally justified, the institution must make a difference. Consequently, a significant aspect of our understanding of an institution's moral justifiability is its instrumental value. Again, this does not require that an institution produce ***the best*** instrumental response, as that may be a matter of reasonable disagreement, but it must make a difference to be a morally viable candidate.

For reasons of clarity, before proceeding, it is important to explain what I mean by political justifiability. One might take this to mean that there is some established procedure with conventional principles and practices that are known and generally followed, political principles and practices that govern the behavior of a relevant set of agents within a defined practical and normative space. This would be a sort of positivist understanding of legitimacy, that it is a social fact that those whose behavior is governed by the institution recognize that their behavior is governed by the rules promulgated by the institution.

This is not, however, what is meant by politically justified here. Rather, I intend a distinctly moral notion, but also a notion that is distinct from what it means for an institution to be morally justified. The fact that both notions are moral in nature may cause some confusion. As such, to avoid any unnecessary confusion, I hope the following distinction suffices. For my purposes, "morally justified" is about substantive moral matters implicated by the challenge of climate change and the alternative responses available to us. "Politically justified", on the other hand, is related to the principles, practices, and organization of the institution itself. One might think of this as a difference between substantive moral concerns and moral values that determine how our political institutions ought to make and implement decisions.

10.4.1 Morally Justified Institutions

With this distinction in mind, to be morally justified, an institution must be able to identify and accommodate the wide range of relevant moral considerations implicated by the challenge of climate change. It is not enough that the institution is able to identify and provide an institutional process for responding to those moral demands directly implicated by the harms of climate change. A morally justified institution must provide an institutional framework that gives the wide range of relevant moral considerations implicated by the challenge of climate change their proper moral weight and consideration within the institution's deliberations about how the international community ought to respond.

An example may help to clarify the distinction. Imagine an institution in which the sole concern is determining how best to respond to the direct harms of climate change. We can even assume, for example, that it is flawless in its ability to identify the best strategy for responding to the problem of rising sea levels and/or extreme weather events caused by climate change. This is not merely a practical success, but a moral one as the best moral strategy for responding to the moral demands directly implicated by the harms of climate change will include considerations of responsibility for the underlying causes of climate change and the differential ability of states to take the necessary actions, in addition to the more specific instrumental demands of effectively responding to the harms themselves.

This, however, would be inadequate. We could also imagine an institution that takes a broader moral view. One in which the moral concerns being addressed are not just those that are directly implicated by the harms of climate change, but ancillary moral matters as well, ones that are indirectly implicated by the challenge of climate change. It would not merely be concerned with the harms of climate change and how best—as a moral matter—to respond to them, it would also address broader concerns of historic (in)justice, human rights, distributive justice, economic justice, ... etc. In other words, a morally justified institution would provide a deliberative process that generates all-things-considered outcomes, and not merely *ceteris paribus* ones.

An additional requirement for an institution to be morally justified, especially when the matter being addressed is as technologically and scientifically complicated as climate change, is that it must be able to distinguish between matters that admit of determinate answers and those that do not. To clarify, responding to climate change in a manner deserving of respect is a wicked problem. Consequently, any institution tasked with identifying the best response to climate change is one that can disaggregate these different types of concerns, giving them their due consideration and evaluating them in a manner appropriate to their nature.

It should be noted that here we find ourselves amidst one of the many points of contact between the elements constitutive of a legitimate institution's practical effectiveness and its moral and political justifiability. We can imagine that any institution tasked with responding to the challenge of climate change will have to simultaneously address moral and political matters alongside matters of science and economics. It is important, as a matter of moral justification (and arguably political as well), that the institution distinguish between the two different types of considerations. Imagine an institution that treats disputes over the anthropogenic nature of climate change the same was as it (properly) treats disputes over claims of historic (in)justice. This is not merely bad science. It is morally problematic because this increases the risk that the policies chosen will be influenced by bad science, that they will be less effective if not harmful.

One may yet wonder why this is a moral matter at all. Simply put, and this highlights a point of contact between moral and political justifiability, if we assume some degree of epistemic humility in our moral judgments, then we must admit the possibility that our moral judgments can be wrong. Perhaps more importantly, we have no objective standard by which we can determine which of two competing reasonable moral views are correct, ours or someone else's. Thus, as a matter of equal concern and respect for the moral worth of others, our institutions should include processes that demonstrate this sort of epistemic humility when it comes to matters of reasonable moral dispute.

The same cannot be said for matters that ***do*** admit of an epistemically accessible determinate answer. There is a significant epistemic difference between matters of scientific dispute and moral dispute. It would be morally problematic to design an institution that treats such matters the same. The fact that with matters of science, for example, there is an epistemically accessible determinate answer to the dispute, means that to treat it the same as a moral issue that does not, fails to recognize that the epistemic humility in relation to matters of moral dispute ought not be assumed in the case of scientific disputes. The moral role of science in a legitimate institution is to ensure that the deliberations being conducted by the institution are based on the best available understanding of relevant empirical matters. If we treat scientific matters the same as we treat moral matters, we run the morally significant risk of pursuing ineffective, if not destructive, policies, thus leading to more harm, a morally worse outcome caused by treating scientific (and other empirical) matters the same as moral matters.

One might wonder if this commits us to moral relativism, in which case there would be larger problems for a project seeking to establish a morally justified institution intended to address a global problem. Though I have discussed my reasons for rejecting moral relativism at length elsewhere,[4] here I am relying on the simple fact that moral pluralism is compatible with universally valid moral norms. Separating our ability to know an answer from whether there is an answer at all, allows for the possibility that just because we ***can't know*** which moral answer is ***the*** correct moral answer does not mean that there is not ***a*** correct moral answer to be had. In addition, for an institution to be morally justified, the range of moral arguments to be considered by the institution are limited to those that are reasonable.[5] This is likely an unsatisfying response for some, but the conceptual point stands.

10.4.2 Politically Justified Institutions

We are left now to consider what is required for an institution to be politically justified. As noted, this is not a value-free understanding of political justifiability, rather it is deeply linked with the substantive moral matters identified. But, to be clear, this understanding of political justification is committed to, what Jeremy Waldron calls, "the circumstances of politics."[6] In short, political institutions and

[4]Kassner (2012).

[5]*Ibid.*

[6]Waldron (1999).

decision-making procedures become relevant when we are faced with a policy question that demands action, but we are not able to come to a consensus view on what ought to be done, and there is no independent and objective basis for determining what the right answer to the policy question is. Thus, to be politically justified, an institution must be able to disaggregate and accommodate different types of questions and include them in ways appropriate to the nature of the question at issue.

This does not mean that, to be justified, a political institution must exclude matters that admit of a determinate answer from their deliberations, rather it merely requires that such matters enter the institution's deliberations in a different way. They should be relied upon in the practical deliberations of the institution when understanding the underlying empirical matter is relevant. In other words, the subjection of a policy question to a political decision-making procedure is appropriate when the question doesn't have a determinate and epistemically accessible answer, but determinate and epistemically accessible answers should be relied upon when such empirical information is relevant to the institution's deliberations.

As an example, imagine two institutions, both tasked with responding to the challenge of climate change. One institution, let's call it the Undifferentiating-Institution (UI), does not differentiate between matters that admit of a determinate answer from those that do not, treating both in the same way within the practical deliberations of the institution. Let us also imagine another institution, let us call it the Differentiating-Institution (DI), in which the deliberative and decision-making procedures differentiate between empirical matters that admit of a determinate answer and those that do not, treating them differently within the practical deliberations of the institution.

With this in mind, let us now consider two questions central to any effort to respond to the challenge of climate change:

1. Is climate change anthropogenic?; and,
2. What should be done about climate change?

The first question is an empirical matter, a matter of science. It is a question that admits of a determinate answer. The second question, however, is not. It is a question of value, though it could be heavily dependent on the answer to the first question.

UI would treat both questions the same, subjecting the scientific question of the anthropogenic nature of climate change to the same political processes to which we would subject the value-laden question of what ought to be done. This would mean that a lack of general consensus amongst the broader population over the anthropogenic nature of climate change would allow for climate skeptics to challenge the scientific consensus that exists that climate change is anthropogenic, possibly resulting in policy choices that are based on bad science and only deepen the crisis we face.

DI, on the other hand, would treat the scientific question differently. The scientific question would be subjected to a form of inquiry and deliberative process appropriate to its use in a political institution. To be specific, and again echoing an earlier point about the appropriate role of empirical matters in the practical deliberations and decision-making procedures of a legitimate institution, the answer to question (1.) would provide a threshold empirical condition that any possible response would have to be able to accommodate. In other words, our answers to such determinate questions would provide epistemic and informational constraints on our political decision-making. Those answers would provide an empirical condition which any viable alternative response would have to be able to accommodate.

Recall, this is not the same as what one might expect if the ultimate goal was knowledge for its own sake. The level of epistemic certainty needed for an answer to an empirical matter to act as a constraint is lower than what one might demand if one were seeking scientific knowledge for its own sake. The standard would, nonetheless, be built around a deliberative process appropriate to understanding matters of science. We would need to know enough about the anthropogenic nature of climate change and with enough (scientific) certainty to aid us in our ability to distinguish between scientifically viable responses from those that are not.

It is equally important that the institution treats political questions in a manner appropriate to their nature as well. It would be deeply problematic, and a threat to the political justifiability of a political institution if moral and political matters, matters that do not admit of a determinate answer, were included in an institutional decision-making process that employed a standard more appropriate for empirical matters that admit of determinate answers. To do so, would be to fail to appreciate the fact that political questions are ones that, to be settled in a politically justified way, need to be settled through a process that respects the moral equality of the individuals affected by the decision, not by a process that depends on the identification of an independently correct moral or value-laden answer that is, by definition, not epistemically accessible. The end result would be unjustified political righteousness, not political justification.

It would also be problematic if determinate empirical answers were presumed to provide answers to political questions. Doing so would render the latent moral and political questions implicated by an empirical answer clandestine. It would mistake scientific understanding for moral correctness. To explain, let us return to question (2.). There one might take the answer to question (1.), that climate change is anthropogenic, to imply specific moral conclusions. For example, that doing nothing is not an option. In the end, that may be true, but the fact that climate change is anthropogenic does not, on its own, imply that the choice to do nothing is excluded from consideration. To do so would be to unjustifiably draw moral conclusions directly from an empirical answer.

From this, we can conclude that a politically justified institution intended to address the challenge of climate change needs to be able to analyze, understand and incorporate empirical information in a manner appropriate to the challenge and its resolution. Perhaps most importantly, this includes the ability to recognize constraints on viable alternative responses implied by such empirical understandings. We also need a deliberative and political decision-making process that addresses only those questions that are properly political.

There is one additional point about the political justifiability of an institution intended to respond to the challenge of climate change that warrants our attention. It is not enough that the institution identifies the relevant empirical information, which is then included in, and acts as a constraint on, the deliberation over political matters within the institution's political decision-making procedures. It is also necessary, that the institution's decisions are implemented. In other words, the structure of the institution, to be politically justified, needs to include, at least, three distinct but interdependent functions.

It is not the case, however, that any single function has priority over the others. For example, the epistemic function is both a source of information for as well as a constraint on the deliberative and executive functions, but it is also informed by and constrained by the deliberative and executive functions. For example, one could imagine a scenario in which the epistemic arm of the institution identifies what is scientifically and technologically feasible, then the deliberative arm offers a novel response, seeking to balance a range of different value-laden concerns. That possible response is then evaluated by the epistemic arm of the institution. Assuming that the response under consideration is viable, it is implemented. The implementation itself then provides those working within the epistemic branch with real world information, that would then serve as the basis for adjustments to the original evaluation of what is scientifically and technologically feasible.

Understanding that it is only appropriate to subject policy questions that are defined by the circumstances of politics to a political deliberative and decision-making procedure is a threshold matter. As is recognition of the fact that the various functions of the institution need to be individuated but also incorporated within a collaborative institutional process. More, however, is needed. In addition to understanding when a political decision-making process becomes relevant, we need an account of what makes a political decision-making process deserving of our respect, especially when we disagree with the ultimate policy or decision reached.

One might be drawn to some form of democratic political decision-making structure, that respect for institutional decision-making must be grounded in decision-making processes that respect principles of majority rule and universal representation, if not direct universal participation. The intuitive appeal of majoritarian democratic political decision-making procedures is easy to understand as there are good reasons to presume that, all other things being equal, political decisions ought to be made through democratic decision-making procedures.

To be specific, if we assume that the matter with which we are concerned is one defined by the circumstances of politics—lack of agreement over a policy choice that is in need of a resolution without an independent standard through which we can determine a substantively correct answer—then the political justifiability of our decision is dependent on the moral nature of the decision-making procedure itself. In which case, political justification is dependent on the institution in question institutionalizing certain procedural values, namely equal concern and respect the moral worth of all those affected by a decision.

What then does it mean to show equal concern and respect for all those affected by a decision? To begin, it requires that we treat each policy decision as an open matter. To privilege one possible choice over another is to treat those in favor of the former as somehow superior to those who favor the latter. In addition, respecting all those affected equally means that each agent is entitled to participate in deliberations about the choices being considered and, ultimately, entitled to have an equal opportunity to influence the outcome of the decision as does every other individual affected by the decision being made. If met, then the outcome—regardless of its substantive correctness—is deserving of respect because it is the result of a political decision-making procedure in which the equal moral worth of all those affected by the choice being made is being respected.

Much more could be said, but this brief explanation of the connection between political justification and certain procedural values should suffice to explain the appeal of majoritarian democratic decision-making procedures. Simply put, to be democratic a procedure must allow for all affected by the decision to participate in the deliberations and decision-making, and it must be a procedure that makes its final decisions, on all but a few matters,[7] based on a majority vote of those who are affected by the decision ensures that the equal moral worth of all who are affected by the decision is respected. As such, all other things being equal, a majoritarian democratic decision-making procedure satisfactorily institutionalizes the procedural values essential to the political justifiability of an institution.

There are, however, at least two reasons to refrain from such unfettered commitment to majoritarian democracy. First, as was noted, a politically justified institution must distinguish between matters that are properly thought of as political and thus justifiably subjected to political decision-making procedures from those that are not. In other words, it would be inappropriate to subject ***all*** questions to a majoritarian democratic decision-making process. Empirical matters that admit of a determinate answer should be addressed by deliberative and decision-making procedures appropriate to the epistemic nature of the matter in question.

The second reason is related to the underlying moral justification for the use of majoritarian democratic decision-making procedures when faced with (properly considered) political questions. If we assume that the matter under consideration and the public discourse around it constitute the circumstances of politics, then arguments that majoritarian democratic decision-making procedures ought to be employed to resolve them tend to be based on the proposition that majoritarian democracy is the best way to show equal concern and respect for all affected by an action or policy-choice.

As noted above, under such arguments, the moral value of the decision-making process is not tied to the substantive correctness of the outcomes reached by the decision-making process being employed; rather, the value of, and basis for the respect owed to, the outcomes produced by majoritarian democratic decision-making procedures depend on extent to which the process respects certain procedural values. To be specific, the equal moral worth of those affected by a choice must be respected. This requires that they be treated fairly and equally; at the very least, that they, their interests and points of view are represented in the decision-making process.

[7]For a discussion of the differences of view, see Kassner (2006), pp. 482–494.

My concern arises when we consider that the values underlying our account of democratic legitimacy are severable from the procedures intended to implement and institutionalize them.[8] For many, majoritarian democratic decision-making procedures are treated as though they were equivalent to the underlying values.[9] But, as I have argued elsewhere, the values are not only severable from their political instantiation, they are also normatively prior to their political instantiation.[10]

The moral basis for the respect owed to majoritarian democratic decision-making procedures is not that they are majoritarian but that they respect the underlying democratic values of equal concern and respect for all those affected by a decision.[11] Consequently, our standard of political justification should be grounded in a demand that an institution show respect and concern for the equal moral worth of all affected by a decision. This may, under normal circumstances, be best accomplished through a majoritarian democratic decision-making procedure, but such procedures are not the same as the underlying values, and the values have normative priority over the procedure.

In addition to such conceptual reasons, and because I am seeking to delineate a standard of political justification for institutions that are practically politically possible, there are a number of pragmatic concerns with the creation of an institution that employs a majoritarian democratic decision-making procedure to settle political questions related to the challenge of climate change that need to be recognized. Here, again, we see a point of contact between the conditions of practical effectiveness and moral and political justifiability. What can be considered to be practically politically possible is both informed and limited by what is needed to be practically effective.

With this in mind, as to the pragmatic reasons, it would be utopian fantasy, due to the global scope of the challenge of climate change, to think that we could construct an institution in which all of those affected by the challenge of climate change and by the choices and actions taken in response would have equal standing as direct participants in the deliberations of the institution. In addition, as a political matter, it is highly unlikely that we would be able to convince states to agree to the construction of a global majoritarian decision-making procedure of any sort. As such, to be practically politically possible, we will need to consider alternative institutional structures that are our best alternatives to the ideal. Our ability to separate the procedural values which justify majoritarian democratic decision-making procedures from the procedures themselves provides us with the conceptual space needed to make this possible.

The question is not whether the institution has majoritarian democratic decision-making procedures, but whether such political decision-making procedures are capable of adequately serving such values. Do the institution's political decision-making procedures adequately respect that an institution's deliberative

---

[8]Kassner (2006).

[9]Waldron (1999).

[10]Kassner (2006).

[11]*Ibid.*

and decision-making processes demonstrate equal respect and concern for the equal moral worth of all who are affected by the choices made and actions taken?

Another challenge to be considered can be found in the concerns raised by Kant in *Perpetual Peace* regarding the moral desirability of global political institutions. Kant pointed out that effective global institutions—for example, one dedicated to providing and protecting international peace—suffered from several moral problems.[12] On the one hand, if such institutions are capable of implementing their policies, then they are likely too strong, posing a threat to freedom and autonomy.[13] On the other hand, if they are weakened, so as to avoid any threats to freedom and autonomy they might pose, then they would likely be ineffective.[14]

I am not convinced that this is an insurmountable moral challenge. First, this is more of a concern for an institution tasked with delivering global peace than one tasked with addressing a challenge like climate change. In the case of the former, the threat to freedom and autonomy comes from the levels of coercion that would be necessary to ensure that parochial political rivalries do not break out into open violence. The threat of climate change is a common threat shared by all political communities, as such there is common interest in finding an effective solution.

Admittedly, when one considers the significant security threats posed by climate change, the difference may not be as stark as I might assume. In fact, there are now a number of philosophers exploring the extent to which the existential threat posed by climate change to some political communities, for example the island nation of Tuvalu, constitutes a just cause for the use of military force.[15] At present, this may largely be an argument with more rhetorical than practical implications, but it is worth taking note of as it points out ethical considerations that help to illustrate the seriousness with which we should be considering the challenge of climate change. Nonetheless, in my considered judgment, even under this line of argument, the security threat posed by climate change only grows if we continue to seek responses that give states discretion over whether and how to respond. In which case, the condition of last resort found in traditional just war theory would mandate that we exhaust all political and diplomatic options first.

Returning to the Kantian objections being considered, and focusing on the possibility that a global institution with the power to effectively implement its policy choices could pose a threat to freedom and autonomy, there are two further considerations that make the concern unwarranted. First, what counts as an effective institution, is one that makes an impact on the practical deliberations of states. In which case, satisfying that condition of practical effectiveness does not necessarily dictate a level of compulsion of the sort Kant thought necessary for a lasting peace.

[12]Kant (2003).

[13]*Ibid.*

[14]*Ibid.*

[15]I want to thank Marcus Hehdahl for making me aware of this emerging discussion amongst military ethicists.

Second, setting aside this minimal standard for practical effectiveness, individual freedom and autonomy is not valuable without limit. If I am using my freedom and/or autonomy to violate your rights or cause you (unjustified) harm, then you (and others) are entitled to interfere with my actions. In short, the value of freedom and autonomy is contingent on how that freedom and autonomy is being used. If, assuming that we have identified through a politically justifiable political decision-making procedure, a response to the challenge of climate change that balances the various moral and political considerations at stake and does so in a way that does not cause any unjustified interference with freedom or autonomy, it would be permissible to use political coercion to implement this response. In short, not all coercion is an unjustified interference with freedom or autonomy.

The last point to be made regarding this response to the Kantian concerns identified relates to the fact that the decision reached is the result of a political decision-making process that shows equal respect and concern for the moral worth of all affected by a decision. This would mean that all those affected by the institution have had their freedom and autonomy properly respected within the deliberative and decision-making process itself. From this it would be appropriate to presume that the decision reached is not in moral conflict with, though there may be a restriction on, individual freedom or autonomy.

One might argue that the assumption upon which this example relies is methodologically problematic. The methodological problem would lie in the proposition that it is problematically utopian, that there would never, in the real world, be an institution with the ability to identify such a morally unassailable response. In which case, one could argue, we have abandoned the non-ideal concern with practical political possibility and embraced ideal political theory. However, even non-ideal theory is grounded in conceptual analysis and argument. Here, the conceptual point of the example is to isolate the consideration of the moral value of the imposition on individual freedom and/or autonomy implicit in the example. The fact that such an imposition ***could*** be morally justified has significant ethical and practical implications. Namely, if the imposition is morally justified, then the imposition itself is not a moral wrong.

In addition, it is important to note the political nature of the assumed institutional deliberations and response. It is not that such deliberations identify the best answer or the most accurate one from a practical and moral perspective. Rather, the deliberations and response are justified against the backdrop of the circumstances of politics. With this understanding of the background conditions in mind, it is not clear that the assumption is quite so utopian. Whether the imposition is politically justified depends on the ability of the institutional deliberative and decision-making procedures to show respect for the equal moral worth of those affected by the decision, not substantive moral correctness of the outcome.

In the end, there are several conditions for political justifiability to be gleaned from this discussion. First, the institution must incorporate deliberative and decision-making procedures appropriate to the nature of the matter being addressed. Questions that admit of determinate answers should be treated differently than value-laden matters that do not. Second, the institution must accommodate the distinct nature of the various functions—epistemic, deliberative and executive—that an institution must serve, but must also provide an institutional framework in which the interdependence of these functions is accommodated. Finally, the institution must respect the equal moral worth of all who are affected by the matter under consideration, the policy choices made, and the actions take.

10.5 Understanding the Structure of a Legitimate Institution

Though extensive, the discussion has not yet answered what it is that makes an institution legitimate. Rather, we have, thus far explicated the elements of practical effectiveness and moral and political justifiability. There are several tasks that remain before we can conclude this discussion. For one, it would be useful to provide an explicit statement of a standard of institutional legitimacy. In addition, and equally important, is a brief discussion of the different functions an institution must serve.

It makes sense to start with the latter as it will have an impact on how we think about our standard of institutional legitimacy. We can begin by recognizing that any institution will likely be required to serve four distinct functions. First, there is an epistemic function. Specifically, any institution must have a process for gathering and appropriately considering evidence and arguments of different sorts. Second, the institution must have a deliberative process in which alternative policy choices will be considered. This would no doubt overlap with the epistemic function, but it is distinct in the sense that the deliberations are about what ought to be done. To clarify, if we think of the first function as a means of gathering information, the second is intended to provide space for debate based on the information gathered. Third, there needs to be a formalized decision-making process. Debate and deliberation for its own sake is not the goal. Fourth and finally, the institution must have some ability to implement its decisions. Whether the institution is practically effective will depend, in large part, on this final function.

In order to understand what would be required of a legitimate institution we must explicate the relationship between the elements of institutional legitimacy discussed above and the four functions just identified. One might conclude, based on the identification of the four functions, that this would be satisfied by an institution that included separate institutional mechanisms dedicated to each function. Simply matching institutional mechanisms to institutional functions is not, however, enough. For one thing, to be legitimate an institution must both separate matters that admit of a determinate answer from those that do not, empirical matters must be distinguished from value-laden ones. But, the deliberative and decision-making procedures employed also rely upon the answers to epistemic questions relevant to the matter at hand. In short, there is tremendous independence amongst the different functions a legitimate institution must incorporate.

Further explaining the standard of institutional legitimacy developed here, the deliberative and decision-making procedures must be appropriate to the role the matter under consideration plays in the institution. As to those matters that are value-laden, they admit of (reasonable) disagreement and do not admit of determinate answers. As such, the deliberative processes employed to consider such matters must show respect for the equal moral worth of all affected by the decision under consideration, but as empirical matters do admit of a determinate answer, the processes and procedures used to identify and consider such empirical matters need not accommodate the procedural values upon which the justification of the political deliberative processes rely.

As to the treatment and role of empirical matters within the deliberative and decision-making processes of a legitimate institution, such empirical matters serve as a constraint on what is possible. Whatever alternative responses are being considered within the practical deliberative and decision-making processes of an institution must not conflict with what is possible. As such, the value-laden considerations being raised and deliberations concerning what responses ought to be pursued will be constrained by our understanding of such empirical matters. It is not enough to assign all epistemic and information-gathering matters to a single functional branch of the institution, rather, such epistemic and information gathering needs to be delegated to multiple interdependent branches of an institution.

There is an important distinction to be acknowledged between information gathering and deliberation. Information gathering—whether dealing with matters that admit of a determinate answer or those that do not—is about collecting, not necessarily evaluating, the relevant information. Deliberation involves the evaluation of the information gathered. When the matters under consideration admit of a determinate answer, the aim of the deliberations is to determine objective validity and empirical possibility. When the matters under consideration do not admit of a determinate answer, the aim of an institution's deliberations is to provide a public political space where those who disagree can exchange ideas and arguments.

It is worth noting, at the risk of being overly repetitive, that the same interdependence that exists within the epistemic and information gathering function between matters that admit of a determinate answer and those value-laden ones that do not, also exists within the deliberative processes addressing such matters. Such value-laden deliberations will be constrained by what is empirically possible.

This interdependence plays out in two ways. On the one hand, the result of those deliberations over empirical matters will act as a prior constraint on what is possible, thus limiting what responses are considered within the political (value-laden) deliberations of the institution. On the other hand, operating under the assumption that the political deliberations will lead to new ideas, it is important that the empirical deliberations operate as a check. In other words, when the political deliberative and decision-making processes result in a specific policy proposals, those outcomes should then be subjected to evaluation by the empirical deliberative process.

It would make little ethical and practical sense to create an institution that had robust deliberative processes without the ability to decide what policy(ies) ought to be pursued. Thus, in addition to the political deliberative processes, a legitimate institution must have a political decision-making process. At the very least, to be politically justified, that process must demonstrate respect for the equal moral worth of all affected by an institution. It should also be connected to the deliberative branches of the institution.

In addition, as the difference between empirical matters and political matters permeates every function that a legitimate institution needs to serve, the epistemic difference between empirical and political matters ought to be respected in the decision-making processes. As noted, the fact that political matters do not admit of determinative answers requires that our understanding of legitimacy in relation to such matters is grounded in the moral requirement that the processes respect the equal moral worth of all affected by a decision. The fact that there are, by definition, determinate answers to empirical questions, however, means that respect for answers to empirical questions depends on independent substantive standards of validity. In which case, what matters is the extent to which the answers satisfy the independent substantive standards of validity relevant to the empirical matter under consideration, not procedural values that treat all views as being worthy of consideration.

From this, we can conclude that a legitimate decision-making process must recognize the same difference between empirical and political matters that have informed our broader understanding of the functional structure of a legitimate institution. In the case of those decisions related to empirical matters, the form such decisions will take will be an acknowledgment of a general consensus amongst those most familiar with the independent substantive standard of validity relevant to the empirical matter at hand. To be deserving of respect, decisions on political matters must be grounded in decision-making processes that respect the equal moral worth of all affected by the decision.

The final structural element to be discussed is likely the most practically difficult. Namely, a legitimate institution needs to implement its policy choice(s). It is arguably the most practically important function of the institution, because without a means of implementing the policy, we are left with "sound and fury; signifying nothing."[16] It is also arguably the most practically difficult because the normative framework of international relations is based on the sovereign independence and equality of states, but, at least in the case of the challenge of climate change, what is needed is an institution that can motivate states to recognize the limits of their sovereign discretion.

We are led then to ask what counts as a legitimate form of policy implementation. Another way to frame the question is in terms of our earlier discussions of both practical effectiveness and moral and political justifiability. In both cases, the legitimacy of an institution depends on the ability of an institution to make a practical difference. What is required is that the practical deliberations of those whose behavior is governed by the institution are impacted by the institution's implementation of its policy(ies). As a moral matter, that difference must be one that improves our ability to achieve the worthy moral goal(s) being pursued. Thus, to

[16]Shakespeare (1606).

be legitimate, the implementation of the policy(ies) chosen should be evaluated based on the extent to which the practical deliberations of those whose behavior is governed by the institution are impacted and the degree to which that improves our ability to achieve the worthy goals of the institution.

Thus, we now have a standard of institutional legitimacy to be used to assess existing institutions and to guide our efforts at institutional design and reform. There is, however, one additional principle that I think is worth including, not because it is independently justified, but because it is a useful rule of thumb for understanding the legitimacy of political institutions. Namely, our evaluation of existing institutions, as well as our suggestions for institutional design and reform, ought to be guided by the principle of subsidiarity.

Subsidiarity is a principle of good governance found in the Catholic philosophic tradition. To abide by the principle of subsidiarity is to seek out the most parochial institution capable of performing the task at hand. There are several reasons for adopting the principle of subsidiarity. For one thing, the legitimacy of majoritarian democratic decision-making procedures is based on procedural values which are well-served by subsidiarity. It is far easier to demonstrate respect for the equal moral worth of all affected by a decision if the deliberations, decisions and/or implementation of a policy are conducted by an institution that is—in relevant respects—epistemically and politically proximate to those affected.

In addition, to be morally legitimate, an institution must be able to provide a deliberative and decision-making process that provides a space for public political reasoning in which competing value-laden claims and considerations are given their proper due. This is more likely to be achieved when the deliberative and decision-making procedures are accessible to those whose views need to be represented.

One might be concerned that a principle that presumes that we ought to make politics more local is a poor one for a global challenge like climate change. This would, however, fail to understand what exactly it is that subsidiarity recommends. It is not that there is something inherently valuable in having more parochial institutions such that epistemic and political proximity to, and control by, those most affected by a decision is an independent value to be weighed against the instrumental value and practical effectiveness of the institution. Rather, it is a *ceteris paribus* principle, at least as I understand it. If there are two institutional designs that are both capable of responding in a practically effective and morally and politically justified way to the challenge under consideration, then—all-other-things-being-equal—we ought to prefer the one that is more epistemically and politically proximate to those affected by the decisions of the institution. If only global institutions are capable of adequately responding to the challenge of climate change, then only global institutions need be considered.

10.6 Conclusion: A Standard of Institutional Legitimacy

We can now state with a sufficient degree of specificity what constitutes institutional legitimacy. A legitimate institution must be practically effective, morally and politically justified, incorporate an institutional structure that includes and differentiates between the different functions the institution needs to serve, and must be consistent with the principle of subsidiarity. Each of these elements has already been discussed in great detail and will not be repeated here. Despite the fact that I am likely to have failed to account for every relevant moral, practical and/or political consideration relevant to institutional legitimacy, the standard provided should allow for the evaluation of alternative institutional models.

References

Kant, Immanuel. 2003. *Perpetual Peace*. Indianapolis: Hackett Publishing.

Kassner, Joshua. 2006. Is Everything Really Up for Grabs?: The Relationship between Democratic Values and a Democratic Process. *Journal of Political Philosophy* 14 (4): 482–494.

Kassner, Joshua J. 2012. *Rwanda and the Moral Obligation of Humanitarian Intervention*. Edinburgh: Edinburgh University Press.

Peter, Fabienne. 2011. *Democratic Legitimacy*. New York: Routledge.

Rawls, John. 1971. *A Theory of Justice*. Cambridge: Harvard University Press.

William Shakespeare, Macbeth (1606).

Waldron, Jeremy. 1999. *Law and Disagreement*. Oxford: Oxford University Press.

Chapter 11
Conclusion: Some Suggestions for Institutional Design and Reform

> I understand the fatalist view that the window for an effective response is rapidly closing. But in the process, I've also arrived at a more hopeful perspective, one rooted in another side of human nature. Properly stoked and channeled, our instincts could help support a different trajectory.[1]

Before taking the final step in the transition from theory to practice, it may be useful to briefly summarize what has been accomplished thus far. The discussion began by recognizing that many of the most pressing moral and practical challenges facing the international community are transnational in nature. The transnational nature of these challenges and the moral imperatives they generate seem to generate a tension between the international community's ability to adequately respond to such challenges while also respecting sovereignty.

This tension is based on an assumption about the normative breadth of the legitimate zone of sovereign discretion. Simply put, it is widely assumed that the new and emerging challenges facing the international community fall within the legitimate zone of sovereign discretion. One significant consequence of rejecting this assumption is that the current debate over the moral nature and limits of state sovereignty needs to be reconsidered and whether the new and emerging challenges facing the international community fall within the legitimate zone of sovereign discretion is an open question.

In response, I argued that the legitimate zone of sovereign discretion is limited both by limits inherent to the notion of sovereignty itself and by the underlying instrumental moral justification for sovereignty. As noted, however, since the nature of the underlying moral justification for sovereignty is an instrumental one, the legitimate zone of sovereign discretion can only be determined challenge by challenge. As such, understanding the moral limits of sovereignty is a piecemeal endeavor. It is this fact about the moral nature and limits of sovereignty that informed the methodological choice to focus on the development of a framework

[1] Robert H. Frank, "In praise of the herd mentality." *The Atlantic* (March 2020), pp. 9–12, 10.

J. J. Kassner, *Climate Change and Sovereignty*, AMINTAPHIL: The Philosophical Foundations of Law and Justice 10, https://doi.org/10.1007/978-3-030-73578-4_11

for assessing whether a challenge facing the international community falls (or ought to fall) within the legitimate zone of sovereign discretion.

Serving as both a limiting and an instructional case, this framework was then applied to the challenge of climate change. After applying the framework, it was determined that the challenge of climate change does not fall within the legitimate zone of sovereign discretion, nor are there good instrumental reasons for extending the sovereign discretion of states to include the challenge of climate change.

One significant implication of the fact that the challenge of climate change lies outside of the legitimate zone of sovereign discretion is that institutions tasked with responding to climate change are not constrained, at least as a moral matter, by sovereignty. Thus, at least when it comes to climate change, the international community has a freer hand in its institutional response. Since our approach to institutional design in response to the challenge of climate change is not fettered by sovereignty, we needed a standard of institutional legitimacy to guide the evaluation, reform and design of institutions.

In response to this need, Chap. 10 developed a standard of institutional legitimacy to be used to critically evaluate existing institutions and to inform our efforts at institutional design and reform should the existing institutions prove inadequate. Thus, in addition to serving as a conclusion to this inquiry, this chapter completes the transition from theory to practice, by first offering a critical assessment of the prevailing international institutional response to climate change, the Paris Climate Agreement; and second, by making some suggestions for institutional design and reform.

11.1 Assessing the Legitimacy of the UNFCCC and the Paris Climate Agreement

The international community has been engaged in various efforts to respond to the challenge of climate change since at least 1992. Then, in Rio De Janeiro, the United Nations Framework Convention on Climate Change (UNFCCC) was agreed upon, going into effect in 1994. The UNFCCC's key provision was a commitment to ensuring that the international community would act to stabilize greenhouse gas emissions to avoid the potential calamitous impacts of climate change.

Though the UNFCCC did not create an institution tasked with responding to climate change, it provided a framework for the development of future conventions and agreements. Ultimately, it led to the creation of both the Kyoto Protocol and the Paris Climate Agreement (PCA). As such, and in light of this history and the efforts that have resulted in the Kyoto Protocol and the PCA, it would be an incredible act of hubris to assume, without argument or assessment, that the extant institutional framework intended to respond to the challenge of climate change is illegitimate. As such, with this history of international action in response to climate change in mind and with a great deal of humility, the first step is a critical assessment of the current institutional framework. Since at present, the international community's response is built around the PCA, the discussion of the legitimacy of the existing institutional response to climate change will focus on the PCA.

The PCA was agreed to in 2015, coming into force in November of 2016. Unlike the Kyoto Protocol which was legally binding and included a top-down approach to fulfilling the objectives of the UNFCCC, the PCA is built around a voluntary bottom-up approach, in which states have a great degree of flexibility over how they choose to respond to the PCA's objectives.[2] For our purposes, the relevant aspects of the PCA and the institutional structure it created are its reliance on Nationally Determined Contributions (NDCs), biennial reporting of greenhouse gas emissions, the global stocktake mechanism, the institutionally recognized right for states to work cooperatively with one another and the implicit reliance of "naming and shaming" to incentivize behavior.

Assessing the institutional legitimacy of the PCA unfolds in two phases. First, each of the elements of the PCA identified above will be assessed on its own. What functional role does each element fill, and what, if anything, are its virtues in light of the understanding of institutional legitimacy developed in Chap. 10? Then, even if those elements are lacking when considered individually, it may be the case that collectively they can satisfy our standard of institutional legitimacy. As such, the second phase will be a more holistic consideration of the institutional legitimacy of the PCA's institutional framework.

11.1.1 Individual Elements of the PCA

Perhaps the most significant aspect of the PCA, at least for present purposes, is its reliance on NDCs. The objective of the PCA is to limits greenhouse gas emissions so as to keep global temperatures (and thus global warming) "well below 2°C above preindustrial levels and pursuing efforts to limit the temperature increase to 1.5°C above pre-industrial levels".[3] One might think of the reliance on NDCs as merely a pragmatic concession to the political realities of the world in which we live—that this is the best we could possibly hope for in a world of sovereign states.

Though I am not entirely unsympathetic to this view of the underlying cause for the adoption of NDCs, under a more charitable view, the justification for the use of NDCs is tied to the fact that different states have different capabilities and are facing different domestic challenges. Thus, grounding the global response in emissions targets that are decided on domestically allows for the wide range of priorities and capabilities being faced by states to be accommodated.

[2]Ad Hoc Working Group on the Durban Platform for Enhanced Action Second Session, part seven. (1 Oct. 2016).

[3]Paris Climate Agreement.

As a point of clarification, in critically assessing the individual elements of the PCA, the goal is not to offer exhaustive criticisms, but to assess with sufficient clarity and convincingness whether each element of the PCA is wanting in some significant respect. As to the reliance of the PCA on NDCs, though they seek to strike a balance between the primary objective of limiting global warming with the priorities and capabilities of individual states, the balance struck is one that fails to adequately appreciate the transnational nature of the challenge of climate change.

This is problematic first because one of the main characteristics of the moral nature of climate change is that it is transnational in a pervasive way. It's causes and effects are transnational, as are the harms of climate change. Consequently, the international community's response to climate change must recognize and respond to the transnational nature of the challenge of climate change, including the harms it causes. NDC's fail to do so. They treat the challenge of climate change as if it were a local problem, as if each state's actions and harms suffered can be separated from that of others.

One might find this to be a harsh critique, arguing that NDCs are the best response we can hope for in a world of sovereign states. That would depend on whether having NDCs is better than not having them. If we don't have good reason to believe that NDCs have improved or can be expected to improve the international community's response to climate change, then this line of defense for the use of NDCs—the central element of the PCA—fails. For the proposition that NDCs are the best alternative available to be true, it must be the case that we can reliably say that having NDCs is better than not. This further assumes that there is somehow a connection between the NDCs and limiting climate change and greenhouse gas emissions.

Regrettably, within the PCA there is no requirement that there be a causal connection between the NDC chosen by a state and the limitation of climate change. In fact, states are given the flexibility and freedom to determine what their contributions will be based on those considerations that ***they determine*** are relevant. Under the standard of institutional legitimacy being employed, this is an alternative that ought not be included because it fails to limit the responses to those that are practically effective. The PCA has identified a goal that is admirable but has built an institutional framework around a mechanism that we have little reason to believe will achieve the goal. In addition, there is ample evidence that the NDCs that have been identified are woefully inadequate as a means to achieving the stated goal of limiting global warming to "well below 2°C above preindustrial levels and pursuing efforts to limit the temperature increase to 1.5°C above pre-industrial levels".[4]

This is especially worrisome since the inclusion of NDCs within the PCA provide the appearance of a meaningful response without an actual impact, rendering the reality of practical impotence hidden beneath the hollow rhetoric of moral action. Rather than face the challenge of climate change with the alarm appropriate to it, we are assuaged, and the appropriate response dulled, by the mistaken belief that

[4]*Ibid.*

something is being done, that we are moving in a positive direction rather than being dragged along with a current that we don't notice.

The next two elements of the PCA to be considered are both informational. The biennial report that each state is required to submit and the global stocktake which occurs every five years serve several valuable purposes. As a principled matter, they increase the transparency surrounding efforts to combat climate change, providing insight into the efforts of individual states. In addition, they enable the international community to hold states accountable if and when, a reporting state fails to meet its NDCs. Finally, the information provided in the global stocktake allows the international community to see just how effective its collective efforts are at achieving its goal of limiting global warming. Such information ***could*** serve as the basis for adjustment and/or reform.

It would be difficult to deny the value of the information provided through these reporting mechanisms, but it is unclear what role such reporting is intended to play in the PCA's institutional structure. As noted, the provision and collection of information related to each state's efforts and the impact that such efforts are having on the goal of limiting global climate change could be used to enhance the ability of the international community to respond to the challenge of climate change. The fact that the PCA is not legally binding and the NDCs are chosen by states themselves, seems to indicate that there is an underlying belief that states will be self-motivated to do their part, provided they can see that others are doing their part. Even if we assume, however, that the participation of states depends on the action of other states, it is equally possible that this will give rise to refusals to act because states see that others are failing to pull their weight. More troubling is the fact that some states may, based on the information provided, freeride on the actions of others.

Further, though not stated explicitly, it is assumed by many that the main motivational driver for compliance with the PCA is a system of "naming and shaming."[5] The underlying idea is that even though the PCA is not legally binding nor is there a coercive enforcement mechanism in which the threat of sanctions can compel states to fulfill their NDC's, states will choose to do so because they can be motivated by the fear of being identified as a bad actor within the international community. Using the publicity surrounding the biennial reporting and global stocktake, those who fail to fulfill their NDCs will be named and shamed into compliance, or at least greater effort.

There seems to be ample evidence that, at least with regards to the PCA, naming and shaming fails to produce tangible effort by those whose action is most necessary. In addition, since the global stocktake is a cataloguing and assessment of the global effort to fulfill the goals of the PCA, it is unlikely to serve as a basis for the identification of failures by individual actors. In which case, much depends on the biennial reports submitted by individual states. And, since the biennial reports are tied to the NDCs, they can be influenced by a range of other considerations and

---

[5]"Paris climate deal: What the agreement means for India and the world," *Hindustan Times (Dec 14, 2015).*

priorities that each state deems appropriate. Consequently, the whole system of naming and shaming can be avoided by committing to an NDC that places a higher priority on domestic goals other than the challenge of climate change. Note that this does not require the state to be freeriding or acting in some other illicit manner, but it certainly leaves open the possibility and greatly weakens the likelihood that naming and shaming will prove effective. In addition, the fact that this all depends on the NDCs is a further indication of a more general problem with the PCA, one that will be discussed below.

Finally, another significant element of the PCA is the explicit provision formally recognizing the ability of states to cooperate with one another in collective efforts to reduce greenhouse gas emissions. This may, at first, appear unnecessary. However, the PCA is intended to achieve its goal of reducing global warming by allocating that responsibility to individual states, thus the significance of the NDCs. Without a formal recognition of the ability of states to act cooperatively and collectively to satisfy their NDCs, it would be unclear if such efforts would count or how they ought to be counted in the biennial reporting or global stocktake. As such, formally recognizing this possibility enables states to address climate change through transnational cooperative efforts, which, as has been argued at length, is what is needed for the international community's response to climate change to be effective.

In light of the previous discussions regarding the transnational nature of the challenge of climate change and the standard of institutional legitimacy from Chap. 10, the flaws with this provision should be apparent. One of the most significant elements of the discussion to this point has been a defense of the proposition that any institution intended to respond to the challenge of climate change ***must be*** transnational, not in an aggregative way, but in a way that addresses the interdependent and interconnected nature of challenge of climate change. Making such action permissible but not required when combined with the nature and function of NDCs within the PCA, there can be little doubt about the inability of these elements to provide a practically effective and morally justifiable response to the challenge of climate change.

11.1.2 Critical Assessment of the PCA As a Whole

Having offered a critical assessment of each of the individual elements of the PCA, a more holistic assessment of the institutional legitimacy of the PCA is in order. To begin, we must ask whether the PCA is practically effective. As a general proposition, this can be answered by looking at the actions taken by the international community since the implementation of the PCA in 2016. The evidence is clear, there is little reason to believe that the international community is on a path to achieving the goals of the PCA.

Nonetheless, we might be doing better with the PCA than we would be without it. As such, it is not enough to rest our rejection of the practical effectiveness of the PCA on the paucity of historical data. Consequently, we should look to what can be reasonably expected of the PCA. The key to understanding the flaw of the PCA is not the fact that it is not legally binding, nor that there is no enforcement mechanism for compelling states to cooperate. Rather, the flaw of the PCA—at least as a matter of practical effectiveness—is the discretionary nature of the NDCs. As has been noted, the challenge of climate change requires that states act in a highly coordinated manner, guided by concerns that transcend political boundaries. NDCs allow states to act individually, giving them the freedom and flexibility to decide for themselves how to balance competing moral claims and priorities, including the choice to set NDCs that are inadequate as a means to combatting climate change. It should be readily apparent that this fails even the most basic understanding of what is required to be practically effective.

One might argue that the appropriate question is not whether the outcome is better, but whether—consistent with the account of minimal practical effectiveness defended above—the practical deliberations of states and the international community are likely to have been altered in such a way as to make achieving the goal of limiting global climate change more likely. It is unclear to me that the reliance on NDCs does anything at all to the practical deliberations of a state. It may make their deliberations more transparent, but I would assume that states are making these evaluations anyway. They are weighing the costs and benefits of pursuing different possible policy choices. Now we have the advantage of seeing how they think climate change fits in to their understanding of the relevant moral and policy landscape, but that falls short of altering the deliberations themselves.

One might, accepting this point, argue that this highlights the significance of the system of naming and shaming—why it may be more effective than was previously acknowledged. To be specific, the argument would run as follows: In identifying a NDC, a state is, at least rhetorically, binding itself to a commitment that can then be used as a basis for criticizing the state's behavior should it fail to meet its NDC. It is reasonable to assume that all states want to be known to be members in good standing within the international community. Consequently, the mere identification of an NDC can alter a state's practical deliberations because it is a self-imposed constraint on unfettered discretion.

This would be true if NDCs were solely concerned with moral considerations surrounding climate change and greenhouse gas emissions. Though, states "should" seek to be progressive in their adjustments to their NDCs, that is not required. In fact, after falling short of meeting their NDC, a state could adjust it so as to be more permissive and they would, under the PCA, be justified in doing so, because such adjustments would be based on the state's evaluation and consideration of the comparative value of other domestic priorities. In which case, the policy of naming and shaming might not be justified by the terms of the PCA itself.

As noted, there are significant moral and political problems with NDCs as a means of achieving the goal of limiting global climate change. In addition, as the entire institutional structure of the PCA is built around NDCs, it would difficult for the institution as a whole to overcome this challenge. Nonetheless, the significance of this disconnect between what is required for an institution to be morally and politically justified and the use of NDCs warrants reiteration. Simply put, the PCA is not morally justifiable because it is unlikely to achieve its intended purpose, thus failing to be justified by its instrumental value. Further, it fails to address the fact that the harms of climate change are negative externalities. Finally, permitting states to identify and choose their own NDCs is to fail to satisfy the basic moral demands that flow from our duty not to harm.

As to the ability of the PCA to satisfy the requirement of political justifiability, if we take it to be the case that to be politically justifiable, an institution must demonstrate respect for the equal moral worth of all affected by its decisions, then it would seem to be the case that the PCA fails. The flaw in the PCA is again tied to its reliance on NDCs. Implicit in the very nature of a NDC is that a state, acting entirely within the rights attributed to it by the PCA, sets its own NDC. That NDC may be based on domestic political and moral considerations that have little to do with climate change or greenhouse gas emissions. Consequently, the NDCs don't account for the fact that one state's decisions about its emissions and the actions it is willing to take impact distant others. Allowing states to prioritize their domestic concerns alongside the transnational concerns related to the challenge of climate change without giving those distant others who are impacted by the state's choices a say in their deliberations, is a failure to respect the equal moral worth of those distant others.

As to the principle of subsidiarity, at first glance the PCA might seems to be consistent with the principle of subsidiarity. To be specific, one might argue that, despite the fact that with regards to all other elements of institutional legitimacy NDCs prove to be problematic, in the case of subsidiarity the institutional choice to give states the freedom and discretion to choose for themselves how they will respond to the global aspiration to limit global climate change would be to assign responsibility to the most politically proximate institution. If we add to that the institutional permission given to states to work cooperatively to meet their NDCs, the argument only seems to become stronger.

This, however, fails to recognize that the principle of subsidiarity does not counsel us to simply seek parochial governance for its own sake, but only when doing so does not undermine our ability to accomplish the institutional task before us. In the present instance, the PCA and its reliance on NDCs is not consistent with the principle of subsidiarity because states are not capable of addressing the challenge of climate change through independently arrived at commitments. What is needed is an institutional structure that not only permits transnational action but demands it. Thus, even though the PCA may at first glance appear to be consistent with the principle of subsidiarity, the fact of the matter is that individual state action is an inadequate response to the challenge of climate change.

In the end, this leads to a general critique of the PCA. From the outset, our discussion of the moral nature and limits of sovereignty has been framed and informed by the fact that the challenge of climate change is a global one. Its causes are global, its harms are global, and its solutions ***must*** be global. One might argue that the PCA is the best we can hope for in the present circumstances. That, however, does not make it legitimate, and a failure to state that and point out its inadequacies explicitly only weakens the necessary resolve to respond in a manner appropriate to the challenge itself.

11.2 Moving Forward: Some Suggestion for Institutional Reform and Design

As much as I would hope to be able to offer details for an alternative institutional framework, that is largely beyond my expertise and ability. However, as I am committed to making the transition from theory to practice, it would be equally unsatisfying to leave this discussion here. As such, the final substantive discussion of the book will provide some suggestions for institutional design and/or reform to be considered by those who possess the expertise and ability I lack. Thus, and with a great deal of humility, the following are some thoughts on the design of legitimate institutions intended to respond to the challenge of climate change.

There are a variety of methodological approaches one could take to understand what a legitimate institution tasked with responding to the challenge of climate change might look like. Here, my suggestions for the design of a legitimate institution tasked with responding to the challenge of climate change begins with a focus on the functional structure of an institution ***capable of responding*** to the challenge of climate change. Thus, at least for the purposes of analysis and explication, priority is given to the condition of practical effectiveness.

This should not, however, be taken to imply that practical effectiveness is more important than moral and political justifiability. Rather, in a legitimate institution, practical effectiveness and political and moral justifiability hang together in an interdependent and mutually supportive functional structure. The exact contours of this structure should become clearer as the discussion proceeds. The point here is merely that the initial focus on practical effectiveness is a methodological choice and nothing more.

With this caveat in mind, to understand whether an institution is practically effective, one must first determine what the goals of the institution are (or ought to be). The answer to this inquiry cannot be assumed as it is a complicated and value-laden matter. Whatever the goal turns ought to be, to be legitimate and worthy of institutional effort it will be constrained by what is technologically and scientifically feasible, but is also in the end a political and moral matter, a value-laden issue that admits of reasonable disagreement. Thus, implicating the condition that the institution's deliberative and decision-making procedures must be politically justifiable.

From this, the first function a legitimate institution intended to address the challenge of climate change must serve is the determination of the goal of the institution itself. It is only when the goal has been identified that we can assess whether the institution is practically effective. Recall, the standard of practical effectiveness here is a relatively minimal one. At the very least, the existence of the institution and the policies it enacts must have an impact on the practical deliberations of those whose actions are governed by the institution. Though it may be obvious to many, it is worth emphasizing that the impact on the practical deliberations of those agents must be in support of the goals the institution is intended to serve.

As to the goal itself, it must be worth pursuing, it must be consistent with the scientific consensus regarding climate change and be consistent with what is technologically feasible. It must also, however, if it is to be morally and politically justified, be decided upon through a deliberative process that gives the relevant empirical evidence its proper role and weight within such deliberations, while also enabling the inclusion of, and deliberation about, the wide range of moral and political considerations relevant to what the goal ought to be. Then, finally, the choice of the goal must be the result of a political decision-making process that respects the equal moral worth of all impacted by the decision itself.

It is worth noting, that one should be attendant to the nature of the goal itself. On the one hand, we should be concerned with the breadth of the goal. Too broad and it would prove largely impotent, not providing a sufficiently clear objective towards which policies and practices can be directed. Too narrow and it would risk excluding moral and practical considerations necessary to a fully justifiable institutional response. In addition, we should be cognizant of the fact that whatever goal is chosen, it is a provisional goal, subject to reevaluation and adjustment, if not abandonment if other more pressing objectives and or considerations present themselves.

With that said, once the goal of the institution has been identified and we are assured that the goal is consistent with the scientific consensus and what is technologically feasible, the political work of policymaking begins. Again, the political deliberative and decision-making procedures are informed and constrained by empirical considerations, but the deliberations over alternative policies and the eventual decisions about what policies are to be enacted are value-laden matters that admit of reasonable disagreement and must be subjected to deliberative and decision-making procedures that are morally and politically justified. Much like the goals of the institution, the policies it chooses to enact must be considered provisional, not in the sense that they are not deserving of our respect, but should the relevant circumstances and/or considerations change, then we should be willing to adjust the institution's policies to meet the new needs.

Once the policies of the institution have been determined, the final phase to be considered is their implementation. When addressing matters of implementation, considerations of subsidiarity are most influential. Recall, that to satisfy the principle of subsidiarity, what is required is that the responsibility for carrying out the policy should be ascribed to the most politically proximate institutions capable of performing the tasks necessary to enact the policy itself. In the case of climate change this will likely be complicated by the fact that the policy will involve measures to address matters of historic and distributive (in)justice, differential capabilities, as well as different approaches to climate change itself.

For example, most approaches to climate change involve some mix of adaptation and mitigation. Adaptation to the threats posed by climate change is a local matter, one in which the actions taken are to the benefit of the local population where such actions are taken. It is also the case, however, that states with the greatest ability to adapt are the ones most responsible, at least causally, for climate change; whereas, those states without the ability to adapt are the ones facing the greatest threats and are least responsible. Mitigation, on the other hand, creates a public good, when one state acts to mitigate its contribution to climate change, others cannot be excluded from the benefits. In thinking about how to implement the policies chosen consistent with the principle of subsidiarity, it is important to keep in mind that it may be wise to disaggregate the elements and measures needed to enact the policy, rather than to seek a one-size-fits-all solution.

11.3 Modeling a Legitimate Institutional Structure

One may find the discussion thus far to be deeply unsatisfying, arguing that we are in need of something more concrete. As such, the following is a model institution that satisfies the requirements of institutional legitimacy. Ideally, one will find the institutional design put forward here compelling despite the fact that is more an impression of what such an institution might look like rather than an explicit design. Alternatively, and somewhat less ambitious, even if it is found to be wanting, the model serves two valuable purposes. It makes the more abstract discussion clearer, and it will hopefully serve the valuable purpose of providing an example around which a fruitful discussion can proceed.

With that said, any discussion of the design of an institution intended to respond to a particular challenge must begin with an understanding of aspects of the challenge that are relevant to a legitimate response to the challenge in question. In the case of climate change, those considerations have been mentioned at various points throughout this discussion. As such, the following is only a brief restatement of those considerations.

First, in terms of causation, climate change is transnational, as are the harms suffered. Second, in addition to the moral concerns directly related to the challenge of climate change, the institution must also consider a broad range of relevant moral matters. For example, a legitimate institution must acknowledge and properly consider matters of historic and distributive justice, the relative capacities of different states to adapt to the threats posed by climate change, the inequity implicit in the fact that the least able to adapt are also the least responsible for climate change, and the impact any particular response to climate change will have on other morally valuable goals and interests.

Third, and finally, we can assume that any legitimate policy is likely to include a mix of mitigation and adaptation. Mitigation has a fundamentally different incentive structure and scope than does adaptation. Consequently, any institution tasked with implementation will need to account for such differences. It is certainly possible that there are additional considerations that need to be taken into account if we are to have a complete understanding of the challenge of climate change, but it should be readily apparent that climate change is a wicked problem, regardless of the perspective through which one is viewing the challenge.

There are three distinct but interrelated elements that any legitimate institution intended to respond to the challenge of climate change must address. First, it is necessary that the institution include some capacity for goal setting. The goal of an institution defines what policies will be pursued and is the basis for assessing the institution's practical effectiveness. Second, having identified what the goals of the institution will be, it is essential that the institution have the capacity for policymaking. The importance of this element should not be overlooked. To be effective, any response to the challenge of climate change will require highly coordinated activity by all relevant actors; consequently, the ability of the institution to meet its goals will require that the policies chosen contribute to a unified and coordinated plan. Third, and finally, the institution must be able to implement the policies chosen in a way that, at the very least, alters the practical deliberations of those whose behavior is governed by the institution in a way that increases the likelihood that the stated goal will be achieved.

11.3.1 Goal Setting

Looking first to the identification of the goals of the institution, we must begin by recognizing that determining what the institution's goal(s) ought to be is likely a matter of considerable controversy. Thus, from the outset, a legitimate institution will need both deliberative and decision-making capacities to determine what the goal(s) of the institution will be. Also, following the principle of subsidiarity, due to the fact that the challenge under consideration is a global one and we have already demonstrated that it is not capable of being adequately addressed through a normative system grounded in sovereignty, the deliberative and decision-making body responsible for determining what the goal(s) of the institution should be, will need to be global in nature.

This does not mean, however, that every aspect of the institution needs to be global. In fact, as I will argue below, the implementation of the policies chosen will likely need to be enacted through regional intergovernmental bodies as well as through the political apparati of states. In the end, the institutional framework being imagined is one in which a range of local, regional and global institutions are used in an interdependent way to pursue the institution's goal(s).

Returning to our discussion of goal setting, the deliberative and decision-making bodies tasked with identifying the institution's goal(s) must have access to the expertise and information relevant to the identification of the institution's goal(s). This epistemic need should be served by various standing committees, each tasked with delivering the needed information or advice. At the very least, there should be a science and technology committee, an ethical committee and a committee on economics. It may also be helpful to have a committee of lawyers and political scientists to offer their expertise when relevant.

What goals ought to be adopted depends on an accurate understanding of the scientific consensus and technological feasibility of the candidate goals proposed, a committee of scientists and technologists should provide an assessment of the existing state of scientific knowledge and technological capability. This assessment should include the current scientific consensus on the causes of climate change, the current and (likely) future impacts of climate change, and the presently available technological responses as well as those that are near term possibilities.

With this information in hand, a deliberative body made up of appointees should begin to discuss what the objective of the institution should be. This is a value-laden political question, and as noted, the empirical understanding of the challenge will not determine a correct answer to what the goal(s) should be but will constrain what goals are worthy of consideration. As to the deliberative body tasked with determining what the goal(s) of the institution ought to be, the appointees to this body are not representatives of the interests of states. Rather, they are representatives of the interests of individuals within specific political communities, thus ensuring that the institution respects the equal moral worth of all affected by the institution's decisions. It is likely the case that there will be extensive overlap between individual and state interests, but they are not the same, and respect for the equal moral worth of all affected demands that we give priority to the former.

In determining what the goal(s) of the institution will be, this deliberative body will be advised by the aforementioned committees. The committees will not merely serve the deliberative body. Instead, the committees will be independent participants in the deliberative process, tasked with offering their critical assessment of and advice regarding matters under consideration that fall within the ambit of their expertise. In addition, they would be responsible for ensuring that all relevant considerations are given their due consideration, which may include raising matters that have yet to be discussed. In other words, part of the charge to those committees would be a responsibility to ensure that the deliberations were sufficiently informed from all relevant perspectives.

The science and technology committee would have an additional role to play after a set of alternative goals had been identified. Any candidate goal to be subjected to a final decision must be one that is consistent with the existing scientific consensus and one that is, broadly speaking, achievable through the use of feasible technological capabilities. Any alternative that passes this threshold could then be subjected to a political decision-making procedure tasked with the final identification of the institution's goal.

One might object that allowing experts to intervene to exclude some alternatives is a violation of the requirement that, to be politically justified, the decision-making procedures employed by an institution must respect the equal moral worth of all affected by a decision. By allowing a committee of experts to intervene and substitute its judgment for that of the political appointees, one could argue that those experts are being treated as superior to others, namely the political appointees and those they represent.

This objection fails to take note of the fact that demonstrating respect for the equal moral worth of all affected by a decision does not require that we treat empirical matters that admit of a determinate answer as if they were open questions. In this case, the limited instances in which a committee of experts can intervene to exclude certain alternatives, are those in which the matters under consideration do admit of a determinate answer. Only those matters that admit of reasonable disagreement and lack a determinate answer are required to be subjected to a political decision-making procedure.

If there is a determinate answer, as is assumed here, then that answer should be given priority over a process that could lead to a goal that contradicts the relevant empirical information available. It is worth pointing out that, though logically possible, it is a near practical impossibility that once the committee on science and technology has excluded those alternatives that are inconsistent with the science or beyond what is technologically feasible, that they have, by a process of elimination, reduced the alternatives to one. It is far more likely that, as a result of, there will remain several alternatives from which the decision-making body can choose.

Notwithstanding this, once the viable alternatives have been identified, it is necessary to decide between them. A choice will still need to be made. Assuming that the remaining disagreements over which goal ought to be chosen are reasonable, what determines the justifiability of the choice is that it is the result of a decision-making process that respects the equal moral worth of all affected by the decision. This should occur through a series of votes amongst the political appointees. In each round the lowest scoring alternative is removed until only two alternatives remain. Then, in the final round of voting, the alternative obtaining a majority of votes should stand as the goal of the institution. Throughout the process and up until the final decision is made, compromise and adjustment should be possible. However, any adjustments to an alternative must be approved by the committee on science and technology.

11.3.2 Policymaking

Having decided upon the institution's goal(s), the next task that must be undertaken is the determination of the policies that will be enacted in pursuit of the institution's goal(s). As a matter of subsidiarity, it remains the case that the matter of policymaking is a global task requiring a global solution. Consequently, the deliberative and decision-making body employed to determine what the goal(s) of the institution ought to be should also serve as the policymaking arm of the institution.

Further, the role of the committees and political appointees remains the same. In deliberating about what policies ought to be undertaken, in addition to responding to specific inquiries by the political appointees, the relevant committees will have the same role and responsibilities of an independent contributor to the deliberative process. In this case, the role of the committee on science and technology is somewhat simplified. They are responsible for ensuring that the policies proposed are consistent with the goal(s) of the institution.

It is the role of the ethics committee that becomes more complicated and, arguably, more important when the focus shifts to policymaking. As specific policies are discussed, those who are trained in ethics will bear the responsibility of ensuring that the moral complexity of each policy alternative is adequately understood. For example, it will be important for there to be an independent and rigorous accounting of the ways in which a particular policy might raise concerns about historic and distributive justice, moral responsibility both past and present and inequity and unfairness in the sharing of burdens. These are but a few of the moral complications that are likely to arise with any policy proposal. The ethics committee is not to decide the political question, but it is responsible for ensuring that the deliberations are adequately morally informed.

It is also important that additional advisory committees be included when the political appointees turn their attention to deliberations directed at policymaking. For example, there will likely be a need for both legal and political scholars, as well as economists. As to the committee on law and politics, they should be responsible for advising the deliberative body of existing international laws and political practices that bear on their policy deliberations. For example, it is necessary that those assessing or comparing different policy options understand the likely impact those policies might have on legal human rights norms and practices. The committee of economists should assess the costs, both intended and unintended, of the various policy alternatives under consideration.

In the end, much like the process governing deliberations about the goal(s) of the institution, it is essential that only those policy choices that pass scientific and technological muster are considered in the final decision-making of the institution. The main difference between the standard of viability employed in the goal-setting phase as compared to the policy-making phase is that the policies chosen are measured against the goals of the institution rather than against a broader scientific background, though the latter remains relevant.

Having identified a range of viable alternative policy proposals and having engaged in informed deliberations and discussions about the moral and economic implications of each, the institution must decide what policies are to be followed in pursuit of the institution's goal(s). We are led again to a decision-making process that needs to respect the equal moral worth of all affected by the decision in question. What ought to occur is the same decision-making process that was used to identify the institution's goal(s), multiple iterations of a majoritarian decision-making process amongst appointees that requires a majority to determine what the policy(ies) of the institution will be. After each iteration the least preferred alternative is excluded from the next round, and any policy adjustments must be approved by the relevant committees.

11.3.3 Implementation

We are left, finally, with the implementation and enactment of the chosen policies. We are assuming that we have a goal that is worth pursuing that it is supported by policies that have both been decided upon by a deliberative process that respects the difference between matters that admit of a determinate answer and those that do not and decision-making procedures that respect the equal moral worth of all affected by the decisions of the institution. Yet, as a practical matter, if the implementation and enactment of the policies decided upon fail to deliver a change, this all amounts to little more than an intellectual exercise.

Thus, when considering how we ought to implement the policies the institution has decided upon, we should be focused on the practical effectiveness of the policies. For the sake of argument, let us assume that the institution's policy prescriptions include a mix of adaptive and mitigating measures. This should be a largely uncontroversial assumption. Climate skeptics might argue for some form of business as usual, but their views will likely have been precluded from our considerations by the committee on science and technology.

One might argue that our response to climate change should be governed by the free market, arguing that adaptations will arise based on the needs of those impacted by climate change and that mitigation will occur in a similar fashion. There will be a demand for technological solutions driven by what people value, rather than by what policymakers decide.

I find this objection to be untenable for a variety of moral and economic reasons. First, as a moral matter, a free-market solution fails to account for the fact that those suffering the most from climate change are both the least responsible and the least able to pay for adaptation or mitigation. Second, those who are most able to pay for adaptation and mitigation are those most responsible. In both cases, there is a serious moral inequity that is being institutionalized rather than rectified by a market solution. Third, climate change is a negative externality caused by industrialization, it's harms should not be borne by those with little or no responsibility for the harm in the first place. Fourth, climate change is a public bad, but in the case of adaptation – since one can exclude others from partaking in the benefits produced – there is little incentive for the haves (who are causally and morally responsible for climate change) to aid the have nots (who are largely neither causally nor morally responsible for climate change). Finally, any effort at mitigation produces a public good, in which case there is little market-based incentive for mitigation efforts because there is no way to exclude others from partaking in the benefits once provided, and there is a weighty incentive to freeride on the efforts of others.

It is important to note, that this is not a rejection of ***market-based*** solutions like an international market set up to allow states to trade carbon credits or any other form of cap-and-trade system. What I am arguing is that there is little reason to believe that the challenge of climate change can be solved by a ***free market*** solution, one in which there is little to no government involvement and the solution evolves from unfettered economic and material interests and interactions of agents.

Notwithstanding this, I have no doubt that thoughtful economists and political theorists would be able to employ carbon-trading schemes to good effect in our efforts to mitigate the harms of climate change.

With that said, returning to our assumption that it is likely that any plausible policy solution will include a mix of mitigation and adaptation, whether their manner of enactment is legitimate will depend greatly on the extent to which the policies chosen are implemented and enacted in a way that satisfies the condition of practical effectiveness and is responsive to the principle of subsidiarity. In now turning to some specific thoughts about the implementation and enactment of any policy that includes adaptation and mitigation, one must recognize that there are salient differences between the two, such that our method of implementation and enactment should treat them differently.

The relevant differences between mitigation and adaptation are to be found in their respective incentive structures and in the scope of the benefit they provide. As to the significance of the incentive structures, this is important to appreciate because it is integral to our ability to implement the policy in a way that is practically effective. More to the point, that difference must be one that guides the agent's behavior in the direction of the goal(s) the institution is intended to serve. Different incentive structures require distinct policy enactments to make a difference in the practical deliberations of an agent.

As to the scope of the benefits provided, efforts at mitigation produce public goods. More importantly, those public goods are transnational. The benefit provided, limiting global climate change, is a benefit from which no one can be excluded. Adaptation, on the other hand, is local in nature. The benefits provided by efforts at adaptation are ones from which others can be excluded. As such, it is possible to pursue adaptation on a more parochial level.

From this brief discussion, we can conclude that implementing mitigation policies needs to be addressed independently of the implementation of adaptation policies. What that implementation will look like is guided by the requirements of practical effectiveness and the principle of subsidiarity. Looking first to the implementation of mitigation policies, due to the fact that mitigation produces a public good, there is, at present, little incentive for any state to act on their own, while there is, in fact, a heavy incentive to freeride on the efforts of others. Thus, the implementation of any mitigation policies will need to generate reasons to be included in the practical deliberations of states in favor of pursuing mitigation.

How might this be done? Due to the incentive to freeride, it is necessary that we look to an institutional structure that transcends states and the state system. One might be drawn to a model of global governance in which states are held to account by an independent political organization, and though I find that appealing for many reasons, it is beyond what can be reasonably expected now or in the near future. Rather, we should leverage existing regional institutions and the sense of legitimacy found in the United Nations.

To be specific, if responsibility for implementing the mitigation policies decided upon by our deliberative and decision-making body is assigned to regional organizations, those organizations would then serve as a mechanism for coordinating the actions of their members, holding them to account through threats of diplomatic isolation and economic alienation as well as naming and shaming. Those regional organizations should also have the formal right to bring forward resolutions before the United Nations Security Council (UNSC) to put further pressure on states failing to meet the demands implicit in the mitigation policies being implemented.

There are at least two concerns that might be raised against this proposal. First, recall that in the PCA, the means by which this was to be accomplished was through a system of "naming and shaming". I objected that, under the PCA, there was little chance that "naming and shaming" would be an effective means of altering the practical deliberations of states. One should not misunderstand this objection. It is not to naming and shaming in general, but to the system of naming and shaming implicit in the PCA. In that case, the reliance on NDCs undermined the possible effectiveness of naming and shaming. Here, since the commitments are predetermined by various political organizations outside of the state, it is more likely that a state could be "named and shamed" for running afoul of those commitments.

The second concern is related to the authority necessary for the UNSC to become involved in the matter. As a general proposition, the UNSC is only permitted to involve itself in the affairs of a state when that state poses a threat to international peace and security.[6] At first glance, this may lead one to conclude that the UNSC lacks the legal authorization needed to involve itself in a matter having to do with climate change since, on its face, that has little to do with international peace and security. I would argue, and there are others who would agree,[7] that the direct existential threat climate change poses to some states and the indirect impact climate change is having on regional stability[8] constitute a grave threat to international peace and security, thus placing climate change within the jurisdictional ambit of the UNSC.

There should be little doubt that if the requisite mitigation policies were implemented in this way, that they would have an impact on the practical deliberations of states. The formal delegation of responsibility to regional organizations with the ability to ostracize and alienate their noncooperating members would generate reasons for states to want to remain members in good standing within such organizations. The further threat of possible UNSC action is an additional reason for action that would weigh heavily in the practical deliberations of states contemplating whether they will cooperate in the effort to mitigate climate change in accordance with the policy being implemented. Should the UNSC refuse to take up the task, it is a condemnation of its own legitimacy, not a fault with the suggested implementation.

Turning our attention to the implementation of adaptation policies, as was noted, unlike mitigation the benefits of adaptation accrue to the community in which such

[6]Charter of the United Nations, Ch. 7.

[7]I want to thank Marcus Hehdal for pointing out this line of argument to me.

[8]"How climate change can fuel wars: Droughts are already making conflict more likely. As the world gets hotter, mayhem could spread." *Economist*. 5/25/2019, Vol. 431 Issue 9144, pp. 58–60.

measures are undertaken. Consequently, it would be reasonable to assume that those communities who are facing the inevitable harms of climate change are already incentivized to pursue adaptation. In addition, due to the fact that the benefits of adaptation accrue to the community in which such actions are undertaken, one might conclude that the policy ought to be implemented by states using their own resources and their own discretion on how best to fulfill the policy demands being placed upon them.

One might find the simplicity of this implementation strategy appealing. At first glance, it even seems to satisfy the requirement of practical effectiveness and be consistent with the principle of subsidiarity. This would, however, fail to appreciate the complicated moral and practical nature of adaptation. As noted, those facing the greatest threats from climate change and thus most in need of adaptation, are also the ones least responsible for climate change and least able to adapt. While those who are most responsible for climate change are, as a general proposition, the one's most able to adapt.

Consequently, if the implementation of a policy of adaptation is to be practically effective and morally justified, it is not enough to rely on the extant incentive to adapt held by those states facing the threats of climate change. What is needed is some way to incentivize transnational funding in support of those states who lack the ability to adapt, especially in those cases where the state in need bears little causal responsibility for the challenge itself. To do otherwise would both fail as a matter of practical effectiveness and moral justifiability.

Similarly, if we consider the principle of subsidiarity, one might argue that, since adaptation occurs at a domestic level, that it is best served by a policy that gives responsibility to states to determine how to adapt. This would, again, fail to understand the fact that, in the case of those states that contributed little to the causes of climate change, the need to adapt is driven by harms caused by the developed world. Consequently, the question of the institutional level at which the policy of adaptation should be implemented is not determined by the fact that the benefits will accrue to the state in which such adaptive measures are taken, rather a broader point of view must be assumed and the funding for adaptative measures must be separated from their implementation.

As such, we are faced with two distinct phases of the implementation of policies that call for adaptation. As to the incentive to adapt to the threats of climate change, little needs to be done to incentivize adaptive behavior, other than to ensure that the funds provided for adaptive efforts are used for that purpose. One could imagine a developing country that is facing a wide range of domestic challenges, using the money provided to address some problem that the local politicians find more important. To protect against such misuse, each state receiving subsidization in support of their adaptation efforts will be required to provide an adaptation plan—including estimated costs—to be reviewed by the funding agency, and once those plans have been approved and funds provided, recipients will be required to report on deliverables indicated in the plan.

It is important to note that the expectations surrounding what adaptive measures should be taken, should be the outcome of a deliberative process in which the state receiving the funds is able to advance arguments related to the broader domestic challenges they are facing. In other words, the funds should reflect a broader range of considerations than just adaptation alone, but once allocated for particular purposes there needs to be some assurance that they will be used for those purposes.

One might object that this depends entirely on obtaining the funds necessary to support such an endeavor. I will not restate the moral argument for why this ***ought*** to happen, even the PCA recognizes that this is a responsibility of the developed world owed to the developing world. Admittedly, understanding the commonsense moral argument underlying the idea that the funding should come from those responsible for climate change in the first place fails to explain how the policy funding such endeavors will be implemented.

Consistent with the principle of subsidiarity, this should be undertaken by a transnational organization. In addition, and from the perspective of compelling recipient states to use the funds in accordance with their approved adaptation plans, the funding organization needs to have control over the funds being distributed and should be shielded from political influence to ensure against its use for purposes other than supporting adaptation the approved adaptation plans.

This still leaves the problem of making a difference in the practical deliberations of those who will be responsible for providing the necessary funds. Though I am quite certain that others will find this suggestion to be the most troubling, in the institutional structure I am envisioning, this would be accomplished by creating some form of global tax or cap-and-trade system in which the revenues generated would go to a transnational funding organization with the specific charge of funding adaptation efforts in developing countries.

As noted, this model has been offered with a great degree of humility and a recognition of the fact that I am offering prescriptions outside of my area of expertise. This fills me with more hope than dread. If I can imagine an institutional structure in which the goals and policies of the institution are morally and politically justifiable and the policies being implemented are practically effective and consistent with the principle of subsidiarity, then I am quite certain that there are others smarter and more able than I am who can do better.

11.4 Conclusion

It is not my intention to summarize the substance of this discussion. Rather, in concluding this book I would like to stress the fact that in my work I am concerned with connecting theory to practice. With that said, I have ambitions related to both, and this final section of the final chapter of this book is dedicated to cataloguing what I hope to have accomplished.

With regards to the theoretical impact of this discussion, I hope to have offered a moral and philosophical argument about the moral nature and limits of state sovereignty that others find convincing; that, having read this book, others who are engaged in this and related debates take my analysis as a foundation for further work in the field. Despite this hope, I suspect that others will find more than ample reason to disagree with the arguments contained herein. I do not see this as an intellectual defeat, especially in relation to the more philosophical aspects of the book, since success in philosophy ought not be found in offering decisive arguments against which no objection can be raised, but in clarifying matters enough so that that new debates will emerge.

As such, in providing a rigorous and, to my mind, novel assessment of the moral nature and limits of state sovereignty, I hope to have aided in the broader philosophical project by clarifying how we ought to think about the moral nature of sovereignty. To be explicit, in providing a framework that improves our understanding of the role state sovereignty ought to play in the normative framework of international relations and that can also be used to assess to what extent sovereignty ought to constrain the international community's response to various new and emerging challenges facing the international community, I hope to have enabled more fruitful debates and discussion about sovereignty.

It is likely that many will disagree with my analysis and assessment of the moral nature and limits of state sovereignty as well as the framework I developed to assess the legitimate zone of sovereign discretion. But, provided that the arguments and frameworks I have provided allow for those who disagree with me to advance their own understanding, then I have provided a useful service to the field by being a foil for those who offer a different view of the matter.

My practical ambitions also range from the aspirational to the realistic. On the more aspirational end, this project was pursued with the intention that my thoughts and analyses regarding the moral nature and limits of state sovereignty and sovereign discretion will inform the design and reform of institutions intended to respond to the challenge of climate change. Ideally, this would result in a meaningful change in the practical deliberations of states. A change that would further result in a legitimate response to the challenge of climate change. Somewhat less ambitious but still highly impactful, would be a future state of affairs in which the analyses and arguments contained herein prove useful to those with the ability to influence policies related to climate change.

Finally, and perhaps most realistic but also significant, is the role this discussion could play in the politics and rhetoric surrounding climate change and many other new and emerging challenges facing the international community. As noted at the outset, much of what motivated this project was the use of sovereignty and sovereign discretion as a moral justification for the behavior of states when considering how they ought to respond to a variety of new and emerging challenges facing the international community. The rigorous understanding of the moral nature and limits of state sovereignty presented here, provides us with a principled basis for holding our political leaders and policy-makers accountable, laying bare the naked interests of power that, in our globalized age, are far too often hidden beneath the high-minded rhetoric of morality.

The manufacturer's authorised representative in the EU is Springer Nature Customer Service Centre GmbH, Europaplatz 3, 69115 Heidelberg, Germany. If you have any concerns regarding our products, please contact ProductSafety@springernature.com

Printed and bound by CPI Group (UK) Ltd, Croydon, CR0 4YY
30/09/2026
02225290-0002